# Mitosis

*Second Edition*

NUMBER XIV OF THE
COLUMBIA BIOLOGICAL SERIES
EDITED AT COLUMBIA UNIVERSITY

# Mitosis

## *The Movements of Chromosomes in Cell Division*

FRANZ SCHRADER

DA COSTA PROFESSOR OF ZOOLOGY, COLUMBIA UNIVERSITY

*Second Edition*

COLUMBIA UNIVERSITY PRESS

New York 1953

COPYRIGHT 1953 COLUMBIA UNIVERSITY PRESS, NEW YORK

*First printing 1944*
*Second printing 1946*
*Third printing 1949*

Published in Great Britain, Canada, India, and Pakistan
by Geoffrey Cumberlege, Oxford University Press,
London, Toronto, Bombay, and Karachi.

MANUFACTURED IN THE UNITED STATES OF AMERICA

## COLUMBIA BIOLOGICAL SERIES

### EDITED AT COLUMBIA UNIVERSITY

#### GENERAL EDITOR

L. C. Dunn, *Professor of Zoology*

#### EDITORIAL BOARD

Hans T. Clarke, *Professor of Biochemistry*
Samuel R. Detwiler, *Professor of Anatomy*
Theodosius Dobzhansky, *Professor of Zoology*
Franz Schrader, *Da Costa Professor of Zoology*

# Acknowledgments

IN the preparation of a monograph like this discussion and criticism come to have a very special benefit. For this I am grateful particularly to Dr. Sally Hughes-Schrader, Professor K. W. Cooper of Princeton University, Dr. H. Ris of Johns Hopkins University, and to my colleagues Professor A. W. Pollister, Professor M. M. Rhoades, Dr. R. Ballentine, and Dr. F. J. Ryan.

FRANZ SCHRADER

*Department of Zoology*
*Columbia University*
*New York City*
*October, 1943*

# Preface to Second Edition

SINCE 1944 there has been a marked increase in the number of publications dealing with mitosis. These have resulted in so extensive a reorientation in many of our views that a treatise on the subject published in the early forties is badly in need of rejuvenation. This second edition represents an effort to bring that about.

In this, as in some other biological fields, the advances of the future will inevitably take a physicochemical direction. It is to be hoped that such mitotic researches will not lose sight of well-established cytological findings which furnish excellent guideposts along the way. Perhaps that is only saying that the time has arrived—regrettable to some of us—when we can no longer pursue our investigations alone. He is a very rare scientist who has a sound working knowledge of more than one of such disciplines as cytology, biochemistry, and physics, and for most of us the necessity of joint research work is obvious.

The last few years have seen some technical progress that is especially useful in work on mitotic problems. New techniques in cytochemistry, especially when used in conjunction with biochemistry, are contributing increasingly to our knowledge of the structure and function of cell constituents, although the results are not yet sufficiently complete or unified to apply directly to the specific problems of the mitotic process. The general adoption of the phase-contrast microscope represents a large stride forward in the study of the living cell, and the modification of the polarization microscope devised by Inoué promises to become at least as useful in the study of mitotic questions. Finally, improvements in the handling of biological materials for study with the electron microscope justify the hope that data on the submicroscopic structure of the delicate objects here involved will become increasingly available.

Since one or two of my reviewers failed to understand it, I

should like to point out again that in this treatise I am dealing with karyokinesis or mitosis in the old (and correct) sense—a division of the nucleus that involves a spindle apparatus. As such it includes meiotic mitosis but not cytokinesis. The justification for this separation of two cell processes, such as it is, can be found in the Introduction.

It is always difficult to give a just measure of the obligation that one has to fellow scientists who are genuinely helpful in an undertaking such as this. To mention the names of Dr. Sally Hughes-Schrader of Columbia University, Professor K. W. Cooper of the University of Rochester, and Professor Cecilie Leuchtenberger of Western Reserve University is therefore a very scant indication of what I owe to them.

FRANZ SCHRADER

*New York City*
*July, 1952*

# Contents

## I. INTRODUCTION    3

## II. STRUCTURE    6

### Living Cells    6

### Fixed Cells    8

#### TERMINOLOGY    10

### The Actuality of Structural Elements in the Spindle    12

#### THE ARGUMENTS AGAINST A "REALITY" OF SPINDLE FIBERS    12

The lack of structure in the living spindle, 12; The effects of fixing fluids, 13; Microdissection, 14; Experiments on tissue cultures, 15

#### THE CASE FOR A "REALITY" OF SPINDLE FIBERS    15

Belar's observations, 15; Centrifuging, 16; Interzonal structures, 17; Direct observation and birefringence, 17

#### CONCLUSION    20

### Nature and Origin of the Spindle Apparatus    20

#### CENTER    20

#### KINETOCHORE    25

#### ASTER    34

#### SPINDLE CONSTITUENTS    35

Continuous fibers, 36; Chromosomal spindle fibers, 39; Interzonal structures, 43

#### CHEMISTRY    47

## III. HYPOTHESES OF MITOSIS    54

### Experimental Analysis    56

### The Period Prior to Metaphase    58

## CONTENTS

| | |
|---|---|
| Prophase Movements | 58 |
| Metaphase Mechanics | 64 |
| Post-Metaphase Movements | 70 |
| Contraction: Pulling | 70 |
| Expansion: Pushing | 75 |
| Variations: Contraction and Expansion | 76 |
| Viscosity and Hydration | 81 |
| Electrostatics | 84 |
| Diffusion | 90 |
| Streaming: Currents | 93 |
| Hydrodynamic Forces | 97 |
| Tactoids | 100 |
| Chromosome Autonomy | 106 |
| **IV. RELATED PROBLEMS** | **113** |
| Resting Stage | 114 |
| Pairing | 115 |
| Telomere | 116 |
| Heteropycnotic Attraction | 117 |
| Kinetochore Attraction | 118 |
| The Nuclear Membrane | 118 |
| The Pre-Metaphase Stretch | 120 |
| **V. CONCLUSION** | **123** |
| **LITERATURE** | **127** |
| **INDEX** | **161** |

# Mitosis
*Second Edition*

# I. Introduction

THE PRESENT TREATISE deals with the mitotic movements of chromosomes. Roughly speaking, the researches of only the last thirty years are considered in detail, the work of the half-century prior to that time being presented only when necessary as a background. The motive for such limitation is a practical one, for the total volume of work that has concerned itself in one way or another with mitosis is very large indeed and far beyond the compass of a monograph of this type. Moreover there exists in the compendium of Wassermann (1929) an extensive consideration of earlier publications which, though presented with his own hypothesis in mind, gives an excellent survey of the field up to that time. For the same reason the chromosome mechanics that pertain more directly to the problems of genetics are treated only briefly, though an effort is made to show that mitotic and genetic problems are indissolubly linked.

If a dispassionate discussion of the subject of mitosis is possible, it is perhaps chiefly due to the fact that our failure to solve most of its problems is so manifest. With rare exceptions we are filled with proper humility—the humility of the open mind. This has not always been so. Since about 1870 there has been a succession of periods in which triumph seemed to stand on the threshold as, first, observers of the living cell, then students of the morphology of the fixed cell, and lastly the physiologists, marshaled the evidence furnished by their different attacks. But it need hardly be pointed out that each of these periods had a corresponding aftermath of disillusion, always accompanied by a new appreciation of the difficulties of the problem.

The present, reawakened interest in the questions of mitosis owes its origin in no small degree to the development of the study of heredity. As the geneticist delves more deeply into the mechanisms that control the behavior of the chromosomes, he is inescapably confronted with the same problems that baffle the cytologist. But the maneuvers of the chromosomes and the complicated apparatus that is involved in their orderly distribution during cell division are equally important in almost every other field of biological research. Development and growth, be they normal or abnormal, are intimately bound to the process of mitosis, and a successful analysis of its basic mechanisms is as important to the student of embryology as to the specialist who is trying to solve the riddle of malignant growth. Similarly the biochemist and physiologist who are concerned with the functions of cells must inevitably be confronted with the series of phenomena that constitutes the mitotic cycle of the individual cell, and a knowledge of its underlying factors is involved in the solution of most of the questions of cell behavior.

Although the most obvious feature of the mitotic process lies in an orderly distribution of chromosomes to the new cells, it must be clear to every biologist that this—the anaphase—is only the culmination of a complicated but orderly series of steps. The preceding telophase, resting stage, prophase, and metaphase represent a highly involved complex of processes. We are accustomed to think of them in rather simple terms: indeed the formation of a nuclear membrane around the telophase group of chromosomes at the end of a division, the evolutions of the chromosomes during resting and prophase, and the metaphase arrangement of chromosomes in an equatorial plate with the simultaneous formation of a spindle, are well enough known in their external aspects. However, the basic significance of all these maneuvers, in so far as it pertains to mitosis, is still very obscure. Indeed, whether the biologist is conscious of it or not, our failure to solve this great problem has been a hindrance to his progress, and it is with the aim of outlining its present status that the present survey has been made.

It has frequently been argued that cell division should be regarded as a whole and that an analysis of one of its aspects to the exclusion of others can never furnish a final solution. In a sense that is of course quite true. The fact nevertheless remains that the elements which participate in the mitotic cycle frequently show a considerable independence of each other. This has long been recognized and was emphasized for instance by Boveri (1897) who showed that the division cycle of the centers proceeds even when cytoplasmic cleavage is prevented by experimental means. Similarly it has been known that the chromosomes may continue to divide under the same circumstances (Wilson 1901, F. R. Lillie 1906) and this independence of chromosomes and cytoplasm has recently been emphasized once more by the differential effects of hydrostatic pressure (Pease 1941, 1946). Finally, the experiments of E. B. Harvey (1936) show that the complete absence of nuclear material does not necessarily make a cleavage of the cytoplasm impossible.

In short there is every likelihood that the behavior of several elements of the cell may be analyzed separately, and indeed it is because of this possibility that we may have hope of a final solution of the problem of mitosis. Without that the complexities of the process are so immense that one might well despair. It is this conviction which serves as a justification for treating here the mitotic behavior of the chromosomes as distinct from the division of the cytoplasm.

Although I have endeavored to be both reserved and fair in presenting the evidence, I have not refrained from expressing opinions wherever they might help to clarify the issue. It must be realized that in the present state of our knowledge of mitosis any opinion whatsoever will infallibly meet with some dissent, which in itself reflects the confusion in which the subject finds itself at the present time.

The word "mitosis" is used by most biologists as an inclusive term to cover any nuclear division that involves a spindle apparatus and the division of chromosomes. In recent years many geneticists have restricted the term to nonmeiotic cells, a usuage which though not correct is eminently practical from their point of view. That, however, leaves them without a general term, for "karyokinesis" is used by few workers. Since the present treatise is not solely concerned with the genetic point of view, "mitosis" will be used in the old, inclusive sense, and the term "meiotic mitosis" applied to the process in the maturing germ cells.

# II. Structure

**LIVING CELLS**

IT APPEARS that division in living cells was observed by both botanists and zoologists early in the nineteenth century. However, the difficulties involved in the study of living cells are attested by the fact that a more or less exact conception of the mitotic spindle was not attained until the decade 1870-80, when observations on fixed cells were drawn upon for comparison (for instance by Schneider 1873, Strasburger 1875 and 1880, Bütschli 1876, and Flemming 1879).

A sweeping generalization concerning the appearance of the living spindle is hardly warranted. Even the chromosomes, which are perhaps more easily discernible than the other elements, are in some species almost invisible. In respect to visibility the difference that is to be observed in closely related species is remarkable. Thus Belar (1930) points out that the chromosomes of the acridid grasshoppers Chorthippus and Rhomaleum are easily seen, whereas in closely related members of the same family such as Trimerotropis they have practically the same index of refraction as the surrounding substances. If, as Belar suggests, this is due to a variation of water content in the chromosomes of different species, it is easily conceivable that other elements of the mitotic figure are subject to similar optical effects. The same may be said concerning the influence of variation in the pH (Yamaha 1935), and studies based on the optical properties of the live cells of a single species can have no general applicability.

Even in optically favorable cells the living spindle is discernible under the ordinary microscope primarily because the chondriosomes and other cytoplasmic elements do not usually enter into it. They thus roughly outline the extent of the spindle body, which itself is rather clear in appearance and, in the vast majority of cases, shows no internal structure (Fig. 1). This is not to deny that in

# STRUCTURE

many instances the spindle substance does differ in its optical properties from other constituents of the cell. In the spermatocytes of some Coccidae the spindle as a whole is clearly visible before the nuclear wall has broken down (Hughes-Schrader and Ris 1941).

The living aster presents fewer such optical difficulties. Its rays are often rather conspicuous, perhaps chiefly because they are outlined by cytoplasmic granules, but sometimes even when the background of cytoplasm is comparatively clear. This differ-

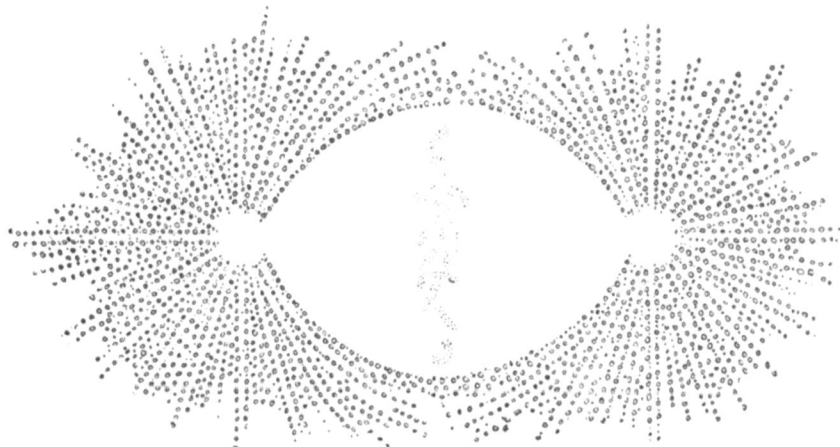

Fig. 1. Metaphase of an early cleavage in the living egg of Rhabditis, a nematode worm.

ence in the visible structure of the living spindle and aster is not without significance, as will appear further on.

Vital dyes that stain the spindle apparatus differentially would of course greatly aid the study of mitosis. But such dyes stain nuclear components only with difficulty and never without detriment to the living cell (Becker 1936, Ries 1938).

That the spindle possesses a certain rigidity has been demonstrated repeatedly. Thus as early as 1905 Foot and Strobell reported that spindles of Allolobophora maintained their form when eggs were punctured and their contents allowed to flow out. Recently, Carlson (1952) has shown that in the living cells of Chortophaga the metaphase spindle, including chromosomes and asters, constitutes a physical entity which can be moved about with a microneedle.[1]

### FIXED CELLS

It will be evident from the foregoing that the orthodox conception of the structure of the spindle is not based on the study of living cells. It is in fixed and stained material that the various constituents make their appearance. The question whether these structural elements exist as such also in living cells is not immediately at issue. It is enough to say that they appear after almost all fixing fluids and that for purposes of further discussion a recognition of their general character is a *sine qua non*.

In one respect the great advances in the cytogenetic research of recent years are not always as helpful for our present purpose as might be wished. The very useful and rapid methods involving the use of aceto-carmine and gentian violet are by no means as favorable for showing the mitotic apparatus as they are for the chromosomes, and many conclusions so based should be viewed with a certain amount of reserve.

Wassermann (1929) differentiated between two main types of spindle, depending on whether centers are present or not, with subdivisions based on the origin of the spindle fibers. This classification was motivated to a considerable extent by the great importance that he attributes to the poles, and is therefore grounded largely on his interpretation of the mitotic mechanism. It is not without value, but the present status of our knowledge of mitotic processes would seem to make it preferable to present a classification without physiological connotations. This has been done for instance by Bleier (1931). I shall modify this classification by emphasizing the spindle fibers which Bleier regarded as artifacts.

On such a basis, just as on that of Bleier, two main types will cover all the forms of spindles ordinarily encountered. The first or "direct" type is one in which the chromosomes are connected

---

[1] Mazia and Dan (1952) were able to separate the entire, intact mitotic apparatus from the egg contents of certain echinoderms, after appropriate chemical treatment. Their method makes possible the gathering of spindles for mass analysis but of course opens the possibility that such spindles do not correspond to normal, untreated spindles in all their properties.

directly with the poles (Fig. 2a). Since such spindles appear to be nuclear in origin they correspond to what Bleier has called nuclear or "Paragenoplastin" spindles. In the second or "indirect" type the chromosome is connected with a continuous fiber which arises independently between the poles. This connection is perhaps established through a chromosomal fiber, but if this reaches the pole it does so only in association with a continuous fiber (Fig. 2b). Since the latter are conceived by Bleier to be cytoplasmic in origin (and in some cases they undoubtedly are), he calls such spindles combination spindles ("zusammengesetzte Paragenoplastin-Cytoplasma

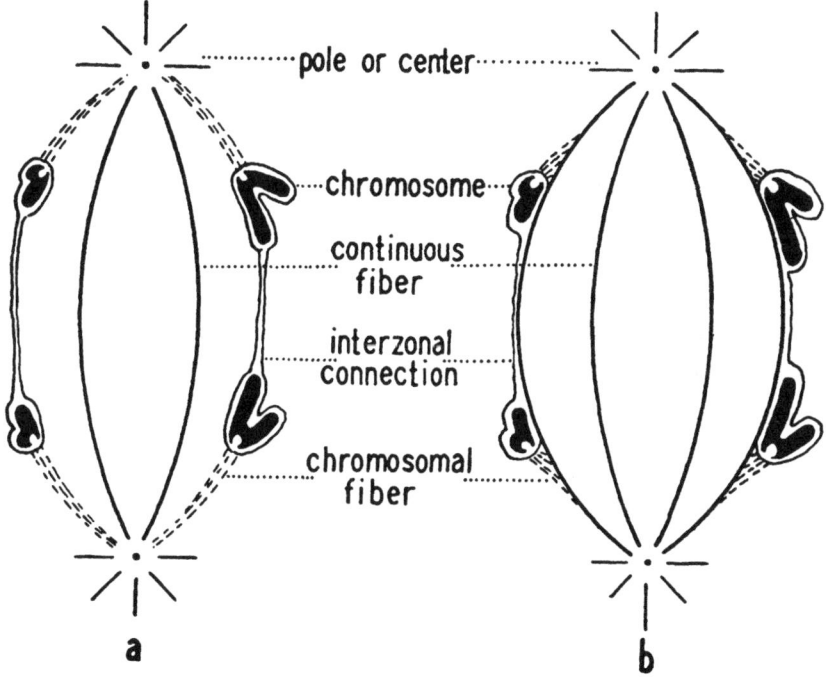

*Fig. 2.* The two main types of spindle as seen in anaphase. a. Direct type; the chromosomal spindle fibers connect the chromosome directly with the pole. b. Indirect type; the chromosomes are connected with continuous fibers.

Spindel"). It will be recognized that such a classification can be subdivided (as is done by Bleier), but for the present purposes the two main types are sufficient; the presence or absence of a discernible center, with or without an aster, does not basically affect it. The criteria used in this classification as well as in others probably have no deep significance. The one here adopted is given only because it seems most practical in discussing mitosis.

TERMINOLOGY. On a purely topographic basis, the region between the center and the chromosomes is called the "half-spindle," while the region which at anaphase and telophase lies between the separating daughter chromosomes is the "interzonal region."

The elements of the mitotic apparatus have been recognized for many years, and redescriptions have frequently been compiled with new terms. Rather than create a new terminology I propose to select arbitrarily those of the multitudinous names which seem rather more familiar than others. For purposes of definition the diagrammatic representation of an anaphase (as shown in Figs. 2a and 2b) is most useful. The components of the various types of spindles may be summed up as follows:

1. "Continuous fiber": fiber connecting the two centers or poles.

2. "Chromosomal fiber" (or half-spindle component): fiber connected with the kinetochore of the chromosome. It may or may not extend to a pole or center.

3. "Interzonal connection": connection between the separating chromosomes as they move apart to opposite poles.

4. "Center": morphologically distinguishable body toward which the spindle elements are oriented. The more noncommittal term "pole" would include cases where there is such centralization but no discernible granular body. As seen in many animals, the center is composed of the spherical centrosome in the middle of which lies a minute body, the centriole. Astral rays may or may not be associated with it. (Wilson 1925).

5. "Chromosome": clearly a complex structure, some or all of whose parts may actively participate in the mitotic process.

A few words on the present cytological conception of its structure must be said here. Although both cytological and genetic researches have made the chromosome the best-known element in the cell a great deal remains to be done. Geitler's able survey (1938) draws attention to some surprising gaps in our knowledge, and no

# STRUCTURE

general statement can be regarded as anything but tentative. The diagram (Fig. 3) here given is a modification of that given by Geitler (1938), with the parts named as follows:

a. "Genonema" (or chromonema) is a threadlike structure in which presumably the genes are carried. It is said to be Feulgen-

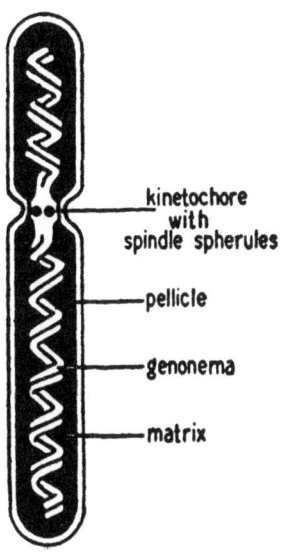

*Fig. 3.* Schematic representation of a chromosome. The number of genonemata or chromonemata is under dispute (two are shown here). Genonemata are shown as relationally coiled, though just before metaphase each is probably coiled independently (parallel coils).

negative, although the chromomeres associated with it are Feulgen-positive. The great majority of cytologists is now convinced that there are at least two chromonemata or genonemata present in each chromosome at anaphase—and probably more.

b. "Matrix" constitutes the main mass of the chromosome and is the constituent that at metaphase characteristically stains with chromatin dyes and is Feulgen-positive. The cytological conception is that of a Feulgen-positive coating around the chromonema. As the loose prophase gyres of the latter become more tightly coiled, the matrix of adjoining gyres becomes confluent and thus imparts the smooth outline to the metaphase chromosome. It is possible that such concepts of chromosome structure will become altered as a result of cytochemical investigations now in progress.

c. "Pellicle" (or sheath)[2] is of all components the most difficult

---

[2] It is unfortunate that several workers (including the present author) have been careless in employing the terms "matrix" and "pellicle" (or sheath), sometimes using one when they meant the other.

to demonstrate. Its presence has been deduced from the behavior of chromosomes during anaphase as well as the nature of interzonal connections (Schrader 1932 and 1935, Upcott 1937b, Carlson 1952), although the hyaline zone described in the fly Lasiopogon by Metz and Nonidez (1924) may represent direct evidence of its existence. The technique reported by Hirschler (1942) seems to make it possible to differentiate a pellicle.

d. "Kinetochore" (or centromere) is the kinetic organ of the chromosome which appears to stand in close structural and physiological relation to the center.

### THE ACTUALITY OF STRUCTURAL ELEMENTS IN THE SPINDLE

As has already been mentioned, there is no doubt that the astral rays as seen in fixed cells correspond very closely to the structure obtaining in living asters. Coagulation and staining may accentuate such rays but they are easily enough seen in many types of living normal cells, and few, if any, workers are inclined to doubt that they exist in such form. But the same cannot be said of fibrous structures in the spindle. The term "fiber" is here used to designate any linear orientation of particles in a longitudinal tract.

*The Arguments Against a "Reality" of Spindle Fibers*

THE LACK OF STRUCTURE IN THE LIVING SPINDLE. The argument that the spindle body is structureless is based on a wide range of evidence. The simplest and most telling point rests on the frequently reported finding that the living spindle appears perfectly homogeneous and that any longitudinal structures that may appear are to be identified with chondriosomes on its exterior. So far as the vast majority of normal cells is concerned this is undoubtedly true. Even with ultra violet light no structures are apparent in such spindles as those of the grasshopper Melanoplus (Lucas and Stark 1931). The lines showing between separating chromosomes in the photographs of grasshoppers cells taken by Wyckoff (1934) appear to be chromatin strands and not spindle fibers. The report of Hughes-Schrader and Ris (1941) that fibers can be seen in the living metaphase spindle of the coccid Steatococcus is not to be doubted, but until such cells are shown to complete a regular mitotic cycle (as these did not) the evidence is affected by the possibility that the conditions were not entirely normal. Indeed it is known that conditions deviating very little from the normal may call out such fibrous appearance, as for instance a slight acidification of the medium (M. Lewis 1923). The best control in such cases would lie in a continuation of the normal mitotic activity of the cells under observation (as in the case of Pediculopsis, p. 14).

The only specific rejoinder that can be made to this adverse evidence is that fibers are not visible under the microscope because their refractive index is very close to that of the spindle substance in which they lie. Admittedly the burden of proof for such a claim lies on the shoulders of its proponents (see p. 18).

THE EFFECTS OF FIXING FLUIDS. Another type of argument against the reality of spindle fibers has been brought forward by many cytologists and especially by adherents of the school of Grégoire (Martens 1929, Robyns 1929, and others). In view of the thoroughness and technical skill that characterizes the work of almost all members of this school, this argument demands serious attention. It rests on the finding that cells in the outer layers of a tissue that are subject to the immediate action of fixing fluids usually show little or no fibrous structure in the spindle. On the other hand, the cells further in the interior where fixation is less perfect are the ones that exhibit the most marked fibrillar conformation. In short the spindle fibers are regarded as an effect of faulty fixation and represent artifacts in the true sense of the word.

There is no gainsaying the conclusions concerning peripheral and interior fixation. However it should be added that the staining is also affected. Thus in the testes of many Amphibia and insects none of the elements of cells near the surface retain very well the dyes that act so vigorously on the deeper-lying cells. Since spindle fibers are without question more difficult to see than the chromosomes and since the latter frequently stain faultily in peripheral cells, it is perhaps not strange that the spindles there appear perfectly homogeneous. No doubt we are dealing with little-known microchemical effects, as Ries and Gersch (1936) and Seki (1933), among others, have pointed out.

Again it has frequently been observed that fixing fluids which cause shrinkage also bring out the spindle fibers. That is true, for instance, of Bouin fixation (see, among others, Belar 1928, p. 701). However, it should be noted that this may well be due to a massing or bunching of very fine, perhaps even submicroscopic, strands already present (Schmitt 1940). The effect is therefore an exaggeration or accentuation of what normally exists.

In short, the absence of fibrous structures in peripheral cells may well be due to technical effects on the staining properties, whereas the coarsely fibrous appearance of spindles results from the secondary action of shrinkage. Neither necessarily constitutes final evidence against the existence of longitudinal structure in the spindle.

MICRODISSECTION. One of the most frequently used arguments against the reality of spindle fibers rests on certain experiments of Chambers (1924). These were performed in the course of a general investigation of the physical properties of the cell and Chambers reports them in a very few sentences. Thus when a microdissection needle was inserted between the chromosomes and the pole, and then moved back and forth, no movement occurred on the part of the chromosomes. It is argued that this shows that the latter can have no fibrous connections with the poles.

The second experiment was to pull a chromosome out of the metaphase plate. Chambers found that no fiber could be seen to adhere to it. He does not describe the conditions of the experiment, but it is not clear why such a fiber should appear after the operation when it is invisible before, unless it is pulled into the cytoplasm. In neither experiment did Chambers take the obvious step of fixing and staining his operated cells to note the effects. But it must be repeated that he was reporting a series of rather random observations which were by no means especially devoted to spindle structure.

Experiments performed more recently with the same type of apparatus by Carlson (1952) led to directly opposed findings. Carlson found that whereas movements of the needle in the general long axis of the spindle could be made freely and with little disturbance, any transverse movement encountered much resistance. His observations clearly indicated some longitudinal connection between the poles and the chromosomes.

EXPERIMENTS ON TISSUE CULTURES. That spindle fibers are artifacts is argued also by M. Lewis (1923) on the basis of careful and well-conducted work on chick tissue cultures. Mrs. Lewis points out that the living cells studied by her show spindle fibers only when the medium has been made acid (pH=4.6) and concludes they are artifacts. There is no reason whatever for doubting these findings, but again, like Belar (1929a) I cannot see in them any decisive point against the existence of fibers in normal cells. The experiment may do no more than to alter optical conditions so as to bring previously invisible structures into view. Indeed Mrs. Lewis reports that exactly the same type of experiment brings out muscle fibrils which also cannot be seen in the cells in normal media. Nevertheless she has no doubt at all concerning the reality of such muscle fibrils.

In short, the case against the reality of spindle fibers may be considered a weighty one, but it is clearly not decisive.

*The Case For a "Reality" of Spindle Fibers*

BELAR'S OBSERVATIONS. In his study (1929a) of mitosis in Chorthippus—a veritable model for cytological investigation—Belar marshaled the evidence for the existence of some kind of longitudinal structure in the living spindle. He pointed out that such evidence throws no light on the form taken by such structure, that indeed it might indicate lathlike strips or lamellae rather than fibers, but that of its existence there can be little doubt. His evidence is indirect and its main points may be summarized as follows:

1. In hypertonic media there is shrinkage of the spindle body, but this shrinkage is much greater in the transverse than in the longitudinal direction.

2. If the spindle body is bent sharply as a result of such treatment, the fixed preparation shows the fibers in the bent region much closer to each other—just as would be the case in a similar configuration of pliable rods.

3. The Brownian movement of particles within the spindle body is decidedly greater in a longitudinal than in a transverse direction.

4. In hypertonic media, shrinkage of the cell and consequent pressure on the spindle frequently induce splits in its substance. These splits always arise in planes parallel to the long axis and never in any other direction. Schaede (1930) arrived at the same conclusion through similar effects produced by centrifuging.

The evidence for longitudinal structure in the spindle is thus very effectively presented, but that it is not completely convincing to Belar's opponents is indicated by the fact that, for instance, Bleier (1931) makes objection even to the last and perhaps strongest of his points—the direction of splitting. Bleier argues that if a body like the spindle is normally under bipolar tension any pressure great enough to produce fracture will result in splits parallel to the axis through the two poles and that no underlying structure is necessary to bring about this result. It must be conceded that the conditions of stress and strain even in a normal untreated spindle are far from understood, but when all of the points made by Belar are taken together they present a strong brief for the reality of longitudinal structure. However, since in the nature of this evidence it cannot finally settle the question, other aspects may be mentioned here also.

CENTRIFUGING. Attempts to throw light on the question through centrifuging have frequently been made. Usually these do not differentiate between chromosomal spindle fibers and interzonal connections and that, as will appear, introduces a source of error. In the present instance these two elements are taken up separately.

Chromosomal fibers: If spindle fibers can be bent in the living cell, they cannot be interpreted as lines of force that lack visual manifestation. Such an argument has frequently been employed, for there is no doubt that spindle fibers often become bent and distorted, either due to natural causes (Bonnevie 1906) or as a result of experimental treatment. This has been shown for instance in the centrifuging experiments of F. R. Lillie (1909), Morgan (1910), Spooner (1911), Andrews (1915), Schaede (1930), Schrader (1934), Beams and King (1936), and Shimamura (1940).

## STRUCTURE

Beyond question their work demonstrates that the chromosomal fibers in preparations fixed after strong centrifuging are often bent, or torn, although offering considerable resistance to such distortion.

The method however does not offer a final proof for the existence of spindle fibers. The objection may well be made that centrifuging, like fixation, quickly induces fibrous coagulation artifacts and that, before centrifuging is stopped, these are distorted. A similar objection holds against the bending of spindle fibers in hypertonic media, and Belar's finding that cells in metaphase rarely recovered from such treatment might well point to such coagulation. Schrader (1934) and Shimamura (1940), working on Crustacea and Lilium respectively, endeavored to meet this argument by showing that controls, simultaneously centrifuged, continued their mitotic activity. However, only the former gave evidence that the mitosis is not even appreciably delayed in the course of the experiment. Nevertheless the somewhat remote possibility of a very brief, reversible coagulation must be reckoned with and the demonstration is not final until this has been eliminated.

INTERZONAL STRUCTURES. In contrast to the chromosomal fibers and the half-spindle in general, the interzonal region is bent with relative ease by centrifuging (Schrader 1934, Shimamura 1940). This was already evident in the reports of earlier workers who, though not differentiating between the various elements of the spindle, show in their illustrations that they too met with this difference. The point is emphasized here because, while the experiments are hardly necessary to establish the reality of interzonal connections (as will be shown below), they do indicate a physical difference between at least some of the elements of the half-spindle and those of the interzonal region (p. 43).

DIRECT OBSERVATION AND BIREFRINGENCE. Despite the consensus of opinion that chromosomal fibers cannot be seen in the living spindle there are now two cases on record that form an exception. Cooper (1941b) found that the living spindles in the cleavage cells of the mite Pediculopsis show longitudinal striations which evidently correspond to the continuous fibers of fixed preparations. Such cells not only completed their mitotic cycle but observations were continued long enough to show that they underwent two additional divisions, sufficient evidence that they were normal. Again, in certain protozoan flagellates, Cleveland (1934, 1935c, 1938b) can see with great clarity the fibers (which he considers astral rays) that run from the center to the nuclear wall, where they are evidently in close rapport with the chromosomes inside the nucleus. Both these cases present rather special conditions however. In Pediculopsis the chromosomes are contained in separate karyomeres which do not break down until late in the cycle, while in the flagellates the nuclear wall never completely disintegrates. Perhaps it is in the delay or absence of nuclear breakdown that the optical conditions favoring visibility of fibers must be sought.

The most generally convincing demonstration that some longitudinal differentiation must exist in the ordinary half-spindle at metaphase rests on the use of polarized light. Schmidt (1936a, b, c, 1937a, b, 1939) reported that under such optical conditions the living spindle shows positive birefringence with respect to its length. Since in the spindle we are certainly dealing with protein molecules and since these are known to react positively to polarized light in their long axis, Schmidt concluded that his observations indicated not only longitudinal structure, but that this structure is due to a parallel orientation of protein molecules. These conclusions have received ample confirmation in the work of Runnström (1936), Pfeiffer (1940), Hughes and Swann (1948), Swann (1951a and b), and Inoué (1952a and 1953). Schmitt (1940), using X-ray diffraction, states it as his conviction that "submicroscopic strands are indeed present in a tenuous, highly solvated lattice."

The finding that fixed spindles exhibit a birefringence identical with that of live spindles led Schmidt further to the belief that fixation induces no fundamental changes and that permanent preparations probably give a faithful picture of the actual conditions.

But Inoué (1951 and 1953), with the aid of an ingenious modification of the polarization microscope, has now given the final answer to this old question. Inoué not only confirms the evidence for birefringence in the spindle but also has demonstrated with convincing clarity that there is fibrous structure in the living mitotic apparatus which in its conformation is very close to what the cytologists have long observed in well-fixed preparations. There are continuous fibers, chromosomal fibers and astral fibers, and to all appearance they are very much alike in their construction.

Although the question of the structures in the interzonal region has not been fully settled as yet (p. 43), the existence of interzonal connections between the ends of chromosomes moving toward opposite poles can be demonstrated in the living cells of several

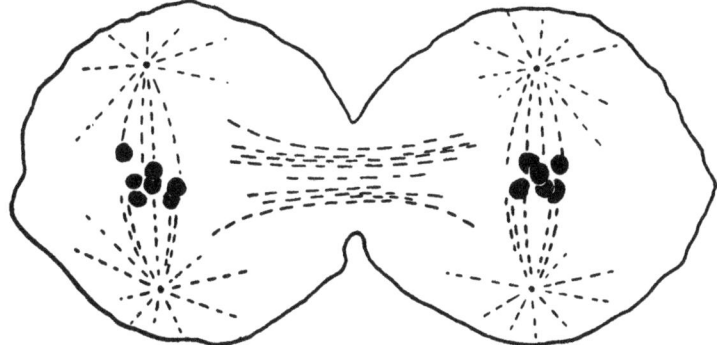

*Fig. 4.* Second spermatocyte division in two sister cells of the bug Pachylis. The interzonal fibers of the preceding mitosis are still present.

species. In some cases, fibrous structures which are not composed solely of interzonal connections become detached from the chromosomes at telophase and persist as independent structures beyond the time when the daughter nuclei have reached the resting stage (Wassermann 1929). In some insects they may still be present (Fig. 4) when the following division is under way (Schrader 1932). They are, in short, structures that have a definite morphological entity, and once formed may exist independently of the forces that are active in the mitotic field. This has also been shown in the centrifuging experiments of Nemec (1927, 1929). It may be remarked that in many cases, long chondriosomal filaments are closely applied to the interzonal region and that the two elements have at times been confused with each other. But no such confusion can affect the present argument. Even in fixing fluids like

Gilson Carnoy, which utterly destroys chondriosomes, the interzonal fibers of some species always persist. Conversely, Hirschler (1935) and Pilawski (1933a and b) have demonstrated that proper fixation and staining will show both elements to exist simultaneously and side by side.

*Conclusion*

The question of the reality of spindle fibers may be regarded as settled. The evidence that has been accumulating for a fibrous structure of the living spindle apparatus has culminated in the convincing findings of Inoué (1952a and 1953). He showed, moreover, that the configuration of such fibrous structure is faithfully preserved in well-fixed material.

The well-established findings that in the living cell astral rays are visible whereas spindle elements usually are not, thus, must be based on optical conditions, as has long ago been suggested. The most obvious basis for this difference is the fact that the astral rays are formed in the cytoplasm whereas spindle fibers are laid down in the substance of the spindle body. As suggested earlier, the difficulty may therefore lie in the fact that the refractive index of the spindle substance and the fibers are in most cases almost identical.

NATURE AND ORIGIN OF THE SPINDLE APPARATUS

The origin and structure of the various elements of the mitotic apparatus would, if correctly elucidated, furnish an answer to many of the questions concerning the movements of chromosomes. The incompleteness of our knowledge of such structure is therefore, in a sense, a measure of our ignorance of the forces of mitosis. The available information is summed up in the following pages.

*Center*

It is obvious that in a typical spindle there is a centralization of certain forces at the two poles of the figure. In many animals the focal point of such activities seems to lie in a morphologically distinguishable granule, the centriole. Earlier observers entertained no doubt of the existence of such a body until Fürst (1898) and

Fischer (1899) suggested that it might be no more than a coagulation artifact. The same argument was advanced once more by Fry in more recent years (1928, 1929a and b) and the question was debated again for several years (Belar 1929c, Wilson 1930, Wilson and Huettner 1931, Pollister 1933, and Sturdivant 1934). Fry's opponents forced him to relinquish the sweeping generalizations with which he had reopened the question and in his 1932 paper on the subject he recognized six different categories of granules that have been reported as centrioles: (1) random granules; (2) focal coagulations; (3) staining artifacts; (4) true centrioles; (5) blepharoplasts; (6) miscellaneous bodies. Thus Fry's final stand is simply that centrioles as morphological entites certainly exist in some cases, but that granules of a different nature are often confused with them. Very few of Fry's opponents would demur at such a statement, which is not dissimilar to one made by Boveri (1900) thirty odd years earlier. The question would thus appear to be settled, and the existence of centrioles as recognizable bodies need no longer be doubted. Fry's subsequent papers (1933, 1936, 1937 and, with Robertson, 1933) did not affect this basic point.

Indeed, the fact that in some forms the centrioles are visible in the living cell would in itself seem to make the claim for fixation or staining artifacts untenable in those particular cases. Reports of such observations in vivo are not rare—for instance, Boveri (1900) in Ascaris, Bresslau (1909) in Mesostoma, Heiderich (1910) in Rana, Johnson (1931) in Oecanthus, Huettner and Rabinowitz (1933) in Drosophila, and Cleveland (1934, 1935c, and 1938b) in various flagellates. Again, a complex but characteristic structure of the centriole, as in the Lepidoptera (Meves 1897c) and certain Orthoptera (Johnson 1931), would seem to rule out its identification with a fixation artifact. Where the involved, centriolar architecture, such as is present in some Protozoa, is concerned, this becomes a certainty (see for instance Cleveland 1938b).

But if no doubt need be entertained that in some organisms centrioles are present as visually discernible bodies, it must also be pointed out that in many other organisms no such demonstration is possible. Thus the conditions in forms like the Echinodermata indicate that the question may not simply revolve about the presence or absence of a granule, for there the focal region of the aster is occupied by a large number of dustlike particles. In addition, certain animals as well as a host of angiosperms show no structure whatever at the spindle poles, and it is a question whether in such cases the substance of the centriole is even more diffuse than in echinoderms or whether we are concerned with a granule beyond the range of microscopic vision. The form of the spindle in certain unorthodox cases would bespeak the first of these alternatives. Thus, there are the maturation figures in the egg of Ascaris where the spindles are truncated; the mitoses in the elongated cells of certain plant tissues (for instance, in the root tips of Drosera) where there is little or no convergence of spindle fibers at all; and finally the spermatocyte divisions of Llaveia and Nautococcus where the spindle actually spreads out at the poles. The best working hypothesis would presume that in all these configurations the material basis for a polarization is diffuse or at least more dispersed than in the orthodox spindle.

There can of course be no dobut that some kind of organizing centers, whatever their condition, are present even in such unusual spindles (Schrader 1932). This fact is especially striking in the eggs of the strepsipteran insect Acroschismus as well as the spermatocytes of Llaveia and Nautococcus already mentioned (Hughes-Schrader 1924, 1931, 1942), in which chromosomal spindle fibers are first laid down helter skelter, but soon thereafter become organized in parallel formation (Fig. 7, p. 36). In other words, the final orientation of the fibers bespeaks the presence of organizing centers on opposite sides of the cell, but their influence does not become manifest until close to metaphase. In Acroschismus the acuminate form of the many individual spindles would argue against the existence of centers in a diffuse condition.

The orderly division of the chromosomes that occurs in all these cases shows that a successful mitotic mechanism can be worked out under a surprisingly great range of polar conditions. Whether the strange mechanism encountered in the spermatocytes of Sciara (Metz 1933), Micromalthus (Scott 1936), Aenoplex (Koonz 1936), Phenacoccus (Hughes-Schrader 1935), Pseudococcus and Gossyparia (Schrader 1923, 1929) belongs under this heading is, however, not certain. In all of these the spindle figures appear to be monocentric, but whether the second center really is absent or whether it is present in an extreme, dispersed condition around the opposite periphery of the cell is not easily answered. Monocentric figures are present without question in many experimental cases among echinoderms and both Metz and Scott so interpret the spindles of Sciara and Micromalthus. The behavior of the chromosomes, puzzling though it be, would seem to support them in this opinion.

That spindle conditions may be affected like any other feature of the organism by changes in the genetic constitution will of course be granted (p. 124). Cases in which the spindle poles have thus been altered are represented for instance by a plant that descended from a Festuca-Lolium hybrid and was found by Darlington and Thomas (1937) to show no polar convergence in its meiotic spindles; and also in Zea plants which, when homozygous for a certain recessive gene, showed a similar failure of polarization (Clark 1940). But these are instances where the affected individuals had not adapted themselves to the change and the absence of a centralizing mechanism resulted in a failure of the telophase chromosomes to aggregate at the poles. As a consequence many separate micronuclei were formed and the whole meiotic process thrown into disorder.

Indeed, the occurrence of animals which show spindles with and without centrioles in the same individual (for instance, the cleavage and meiotic divisions in many insect eggs) would indicate rather special adaptive conditions.[3] For it must be remembered that the existence of normally functioning spindles without centers does not prove the centriole to be merely a nonessential manifestation of underlying forces. In cases where a centriole is normally present, its absence is closely correlated with a loss of polarization, as is shown in many experiments (Painter 1918).

The discussions on the reality of the centriole have only confirmed the older conclusions concerning its mode of origin. In

[3] The case of the maturation in the Asplanchna egg (Storch 1924) would seem to suggest that even when a center, or at any rate an aster, is present, it may show no relation to the spindle that is formed. However, so singular a condition calls for confirmation.

some forms the centriole is a permanent cell organ, and new ones arise normally only through the division of preexisting centrioles, as was argued for Ascaris by van Beneden and Neyt (1887) as well as by Boveri (1887). But it is equally well established that in at least some species new centers of mitotic activity can also arise *de novo* (cytasters); the many early demonstrations of that fact find full confirmation in the experiments of Fry (1925) and Tharaldsen (1926). What is not so clear is whether these new cytasters contain normal centrioles. Unfortunately, the echinoderms on which most such experiments have been performed are characterized by a pluricorpuscular centriole. Nor are the beautiful experiments of Yatsu (1905) decisive on this point; for, though the normal Cerebratulus egg has a sharply defined centriole, the cytasters called out by chemical treatment show a number of fine granules at the center. That cytasters of this sort have many of the properties of the normal division center is attested by the fact that they may bring about a cleavage of the cytoplasm in a small percentage of cases. Nevertheless such an induced formation of cytasters does not in itself establish a *de novo* origin of centrioles. Instead it indicates that the all-important mechanism that resides at the poles of the mitotic figure can be duplicated—at least partially. And thereby is given the possibility for one type of analysis of what underlies the activities of the normal centers.

In this conjunction, mention may be made of the fact that in certain plants a chromosome that has become separated from the spindle may itself become the midpoint of a miniature spindle (see for instance Belling 1927 and Belling and Blakeslee 1924). Whether such a chromosome can induce the formation of miniature poles or whether it carried with it both spindle and polarizing substances is of course hard to say. The case poses the same questions as does Acroschismus, where an individual acuminate spindle is formed in conjunction with each chromosome (Fig. 7). It is likely that we are dealing with special conditions which are not necessarily linked with the displacement of chromosomes from the regular spindle and argue for the possibility that in some species the individual kinetochore is capable of orienting the adjacent protein structure without a collaboration with the center. This in turn bespeaks the fact that the organizing powers are not the same in all kinetochores. Certain it is that not all lost chromosomes form such miniature spindles.

In considering the part played by the centers in mitosis it should be kept in mind that in a great many instances their influence may manifest itself before the metaphase. Thus there is in some species a marked deformation of the prophase nucleus in response to such activities of the centers (Schrader 1947a), and in a host of cases there is a reaction of the chromosomes to centriolar forces long before the nuclear membrane has broken down. The "bouquet" formation of leptotene or pachytene chromosomes is a case in point, and the obvious responses of diakinetic chromosomes to the movements of the centers outside of the nucleus, as in Anisolabis (Schrader 1941a), are good evidence for the fact that the centers play an active role throughout the entire mitotic cycle (Figs. 12 and 14).

*Kinetochore*

An element of fundamental importance in the movements of chromosomes is the kinetochore or centromere. Only in recent years has it received the attention that it merits, although it was reported by Metzner in 1894 and described again by Nawaschin in 1912 and Sakamura in 1920. Biological nomenclature has been at its worst in dealing with it and over 27 terms have been applied to it in the course of the years.[4]

[4] There is no point in keeping alive the many names that have been applied to the kinetochore. The following list of 27 terms does not attempt to give minor modifications, but shows quite adequately the length to which biological nomenclature may go:

| | | |
|---|---|---|
| Achromatic region | Guiding granule | Polar granule |
| Achromite | (Leitkörperchen) | Proximal granule |
| Attachment body | Insertion region | (granule proximale) |
| Attachment chromomere | Insertion gap | Separation region |
| Attachment constriction | (Insertionslücke) | (Trennstelle) |
| Attachment locus | Joint | Spindle fiber attachment |
| Attachment point | (Gelenk) | Spindle fiber constriction |
| Attachment region | Kinetic body | Spindle fiber insertion |
| Centric constriction | Kinetic constriction | Spindle fiber locus |
| Centromere | Kinetochore | Traction cone |
| Commissure | Primary constriction | |

The difficulties in the study of the kinetochore are considerable and are founded in part on its small size. Darlington (1937) estimates its size at 0.2 μ, but Matsuura (1941) in the large chromosomes of Trillium at 3 μ, a discrepancy possibly due to the fact that Darlington measured only the spherule, as well as the fact that no adequate technique has yet been evolved to differentiate it from the rest of the chromosome structure (Schrader 1939). Many of the conclusions concerning the behavior of the kinetochore are based on the configuration of the chromosome as a whole. During middle and late anaphase the kinetic region usually takes the lead on the way toward the pole and in long chromosomes the arms often appear to be "dragged" behind. In such cases the conflexion or "commissure" of the chromosome indicates the location of the kinetochore, even though the latter may not itself be visible (White 1935).

Morphologically, it usually appears at somatic metaphase and late prophase as an uncolored or faintly stained region in the chromosome that has long been known to botanists as the primary constriction. Despite its frequent failure to take a stain, it quite clearly is not a gap in the chromosome structure and in such plants as Zea the pachytene chromosome show it as a distinct body, more or less ovoid in shape, and rather larger in diameter than the rest of the chromosome thread (see for instance McClintock 1933). This kinetochore, though of considerable size, appears to be homogenous and structureless, which may or may not be due to the acetocarmine method employed in the large volume of cytogenetic work in Zea. In certain Amphibia and especially Amphiuma, a second element—the spindle spherule or kinosome—appears in this commissural region. This is differentiated by a greater staining capacity (Schrader 1936, 1939) and it is often difficult to tell whether it is this body or the whole kinetochore that was figured in such cases as Chorthippus (Belar 1929a), Bellevallia, Crepis, Najas (Trankowsky 1930), Bufo (Minouchi and Iriki 1931), Allium (Koslov 1937), and Lilium (Iwata 1940).

The tiny globules which terminate the fine threads emanating from the kinetic regions of anaphase chromosomes are probably to be identified as spindle spherules. In this form they characterize the first meiotic mitosis of many plants like Tradescantia (Nebel

1935, Schrader 1939, Propach 1940), Trillium (Huskins and Smith 1935), and Lilium (Iwata 1940), and their marked appearance at this division may well be correlated with the resistance offered to separation by the chromosomes. Almost certainly the delicate, knobbed threads described by Cleveland (1934, 1938b) and Grassé (1939) in certain flagellates belong in this category. The chromatic thickenings frequently described in acridid Orthoptera and called polar granules by Pinney (1908) cannot, however, be so identified. In many cases these seem to represent terminal regions of the chromosome that are heteropycnotic and though they may of course include the kinetochore, that is manifestly not possible where both ends are so characterized.

It already has been pointed out by Metzner (1894) and confirmed by Schrader (1936) that in certain Amphibia the kinetochore is stained differentially by chondriosomal staining methods, taking the color of chondriosomes rather than of chromosomes. This is true also of the scorpion Opisthacanthus (Wilson 1931, Schrader 1936) and of the mouse (Schrader 1936, Benoit and Kehl 1939). In Amphibia a careful extraction of the dye will decrease the coloring of the main mass of the kinetochore and leave only the spherule thus stained. Schaede (1936, 1937) reported that in the plants Allium, Cephalaria, Picea, Impatiens, Galtonia, and Scorzonera, the kinetochore can similarly be differentiated from the rest of the chromosome because in contrast to the latter it is Feulgen-negative. However Propach (1940), Iwata (1940), Coleman (1940), and Lima-De-Faria (1950) did not find this so in various plants and stated that the knobbed fine threads of the first anaphase are Feulgen-positive just like the chromosome body. If this is correct, we are confronted with a puzzling difference between animal and plant kinetochores but it may be pointed out that in such minute objects the stain of the Feulgen reaction is not always very decisive.

It is clear from the foregoing that a consideration of the detailed construction of the kinetochore must at present stand largely on hypothetical grounds. The indications are strong that we are dealing with a compound structure, but just how this is related to the rest of the chromosome is still an open question.

Nebel (1939) and Darlington (1939a) content themselves with

postulating a liquid body which contains micelles oriented parallel to the long axis of the chromosome (Fig. 5c). This is an assumption derived largely from the hypothetical considerations of certain abnormal cases of chromosome behavior.

Sharp (1934) and Schrader (1939) suggested that the spindle spherule is attached to the genonema which extends through the

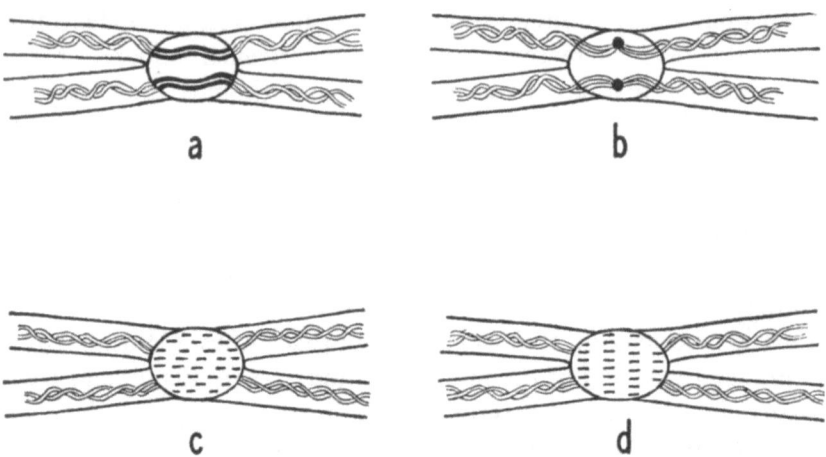

*Fig. 5.* Hypothetical structure of the kinetochore in a chromosome with two pairs of chromonemata. The kinetochore is shown as an ellipsoid containing the essential kinetic material in the form of: a. a regionally modified portion of the chromonema (Matsuura 1941); b. a special body (spindle spherule) connected with each pair of chromonemata (Sharpe 1934, Schrader 1939); c. a number of oriented micellae (Darlington 1939, Nebel 1939); d. an arrangement of micelles facilitating a transverse break of the kinetochore (misdivision) instead of the normal, longitudinal division (Nebel 1939).

kinetochore as well as the chromosome (Fig. 5b). The tensile forces involved in anaphasic movement may pull the spherule out of the kinetochore, stretching the adjacent genonema in so doing. This condition would be represented in the delicate, knobbed threads seen in certain plants already mentioned. Lima-De-Faria (1949a and b) as well as Tjio and Levan (1950) recognized a similar structure in certain plants but described the individual spindle spherule as being double, with its two components separated by a short thread. In this more complex structure Östergren (1951) sought the reason for the protrusion of the kinetic region during certain divisions but Cooper (1951) could find no support for such an interpretation in the insect Boreus.

Matsuura (1941) is inclined to make an interpretation similar to Sharp and Schrader but doubts the existence of a spherule. On the basis of his beautiful preparations of Trillium he concludes that through the commissural substance (his kinetochore matrix) there extends a regionally modified portion of the genonema (his kinetonema). The latter is the permanent component of the kinetochore and what others describe as a spherule is perhaps only a coagulation of its substance (Fig. 5a).

Carothers (1936) departs from these general conceptions of the kinetochore as a highly specialized element. She believes that it is nothing more than a temporarily modified chromomere and that as a result of changes in the timing of condensation other chromomeres may take over its functions.

If the structural connection between the kinetochore and the body of the chromosome is thus a matter of conjecture, its relation to the spindle is less dubious. There is general agreement that the chromosomal spindle fiber that extends between the chromosome and the pole is centered in the kinetochore. In Amphiuma, which offers especially favorable conditions, the chromosomal fibers are connected specifically with the spindle spherule rather than more generally with the whole kinetochore (Schrader 1939). The fact that the individual chromosomal fiber is in many, probably most, cases composed of several fine fibrillae (Fig. 9, p. 42) suggests that the spherule is itself an aggregate of smaller elements, each of which is active in the formation of a fibrilla. The structure of the spherule is thus perhaps similar to the top of a salt shaker or sprinkling can, where the holes represent the component elements of the spherule.

Be that as it may, there is no longer any doubt that the kinetochore plays a major role in the mitotic movements of the chromosome. This has for some years been recognized by cytogeneticists who have found that a chromosome deprived of its kinetochore behaves erratically and soon becomes lost. With this in mind, Upcott (1937), Koller (1938), and Darlington (1939a) made some highly ingenious interpretations of the behavior of certain lagging chromosomes in the meiosis of Tradescantia, Tulipa, and Fritillaria.

According to them, in these cases at beginning anaphase the kinetochore is not ready for the division and as a result undergoes "misdivision." Briefly, this occurs when the kinetochore, instead of dividing normally in the long axis of the chromosome, is severed at some angle more or less transverse to this axis. As a result, there arise: fragment kinetochores to which both short arms (or both long arms) of the two chromatids are attached; or two kinetochore fragments with three arms and one arm, respectively; or akinetic chromosome arms; or, finally, kinetochores freed entirely from the chromosome body. The possibility that the kinetochore is thus responsible for certain abnormalities had been broached earlier by Nishiyama (1931), who did not, however, pursue the subject. Darlington's (1939a) and Nebel's (1939) explanations of misdivision rest on their conception of kinetochore structure. Darlington simply postulates that if the component micelles (his centrogenes) have failed to divide in preparation for anaphase, the kinetochore may be cleaved at an abnormal angle; Nebel imagines a regular cycle in the orientation of the micelles, at only one point of which a normal division can occur (Figs. 5c and d). The speculative nature of both of these suggestions is self-evident.

Darlington's cytological evidence for his conclusions concerning a faulty division of the kinetochore is not necessarily valid for all species. Thus the assumption that akinetic chromosomes cease moving does not hold for the Orthoptera studied by Carlson (1938a), who found that they frequently reach the poles. However there is no doubt that the kinetochore may "misdivide," for by 1932 McClintock had already presented evidence that this may be induced in Zea and that the two resulting fragments with the attached chromosome regions may continue to function normally. Again, "naked" kinetochores freed from their chromosomes, appear to exist regularly in the meiosis of the oligopyrene sperms in certain molluscs, as the Pollisters (1939, 1943) have shown. Independent evidence for the occurrence of functional abnormalities in the kinetochore is thus available.

The work of Rhoades (1940) is of considerable importance in a consideration of the structure and physiology of the kinetochore. He found that a Zea chromosome that has lost one of its arms and thus has a terminal kinetochore may behave more or less normally but will eventually become lost. This would seem to support the contention that the kinetochore is never terminally placed in the normal chromosome, but that, as Nawaschin (1912) first argued, all chromosomes have two arms. This generalization has so far found justification in most cases subjected to careful analysis, and even the dotlike chromosome IV of *Drosophila melanogaster* conforms to this rule (Kaufmann 1934, Griffen and Stone 1940). Very likely we are here touching on the special character of the terminal regions of chromosomes, the importance of which is now realized.

However, in a few instances, the evidence supports the claim for a truly terminal kinetochore, as, for instance, in certain strains of wheat (Love 1943) and in the grass Phleum (Ellerström and Tjio 1950). According to Cleveland (1949a) the large chromosomes of the protozoan Holomastigotoides show the kinetochore to be terminal without a question. In view of such reports, the time is not yet ripe for any generalization concerning the position of the kinetochore.

It is also pertinent to our inquiry that the evidence indicates cyclical alterations in the kinetochore. Thus in Amphiuma (Schrader 1936, 1939) the kinetochore is evidently more diffuse in the late prophase than at metaphase, and it is probable that the commissural region is not present as such until close to the last-named phase. At that time the kinetochore is in most organisms a strictly localized element of the chromosome. However indications are not lacking that in certain groups of animals the conditions of the kinetochore do not conform to this conception. That is patently true in certain nematodes where large multiple or "Sammel" chromosomes are formed by a linear union of many tiny chromosomes. The latter retain their kinetochores and hence chromosomes such as are found in the germ line of Ascaris are characterized by a multiple kinetochore (Painter and Stone 1935, Schrader 1935, White 1936). But though the behavior of such chromosomes poses certain questions that have not yet been answered, it conforms in general to what might be expected from our knowledge of the individual kinetochore. The difficulties are greater however in the Hemiptera, where as was pointed out several years ago (Schrader 1935) the behavior of the chromosomes suggests that the kinetochore is not localized in a certain region but is spread or diffused over the whole length of the chromosome. Clear-cut evidence for such a condition has been brought forward by Hughes-Schrader and Ris (1941) and Ris (1942), who fragmented the long chromosomes of Steatococcus and Tamalia respectively by means of X rays and found that even very small pieces continued to divide in the normal manner at the normal rate for a long time.

The existence of a diffuse kinetochore in the plant Luzula was more recently established in a similar way by Malheiros, De Castro, and Camara (1947), De Castro, Camara, and Malheiros (1949) and De Castro (1950).

In neither case could any morphological indications be found for a multiple or polykinetic condition such as is present in Ascaris, and it is therefore likely that such a diffuse kinetochore exists as a series of units so small that the individual is not recognizable. This possibility is not farfetched, for even the spherules of such localized kinetochores as those of Amphibia seems to be composed of smaller functional units, as is indicated by the fact that each chromosomal spindle fiber comprises a number of very fine fibrillae (p. 41). If such constituent, kinetic elements are spread along the entire side of a chromosome, the kinetochore may well be called diffuse.

In short, whether to call a kinetochore diffuse or polykinetic would then seem to hinge on the size of the component, kinetic particles. And of this we know nothing as yet. The disposition of such a kinetochore with respect to the genonema is however a puzzling question. Certainly the latter at metaphase is in the form of a coil, and, if the kinetochore substance is associated intimately with this, it would not be confined to the poleward regions of the chromosomes. Obviously we are dealing with conditions that are still very obscure, but it may be noted that a diffuse kinetochore may have its parallel in the condition of certain centrioles (p. 23).

Such departures from the more generally accepted structure may not be as uncommon as we have tacitly believed. Some evidence is available that it may also occur in certain Acaridae

(Cooper 1939), some Lepidoptera (Federley 1943, 1945), a few Scorpionidae (Piza 1939 to 1950), and possibly in the Odonata, where the peculiar findings of Oksala (1943, 1944, 1945) would seem to meet with a natural explanation if such a kinetochore were assumed to be present (see for instance Schrader 1947b, and Hughes-Schrader 1948a). The cytological conditions in the scorpion Tityus strongly suggest a diffuse kinetochore though, contrary evidence notwithstanding, Piza has postulated in a long series of publications (the most recent in 1950) that in that species each chromosome has two localized kinetochores—one at each end. However the evidence based on X-ray experiments brought forward by Rhoades and Kerr (1949) furnishes a decisive argument in favor of a diffuse kinetochore.

The discovery of a subsidiary kinetochore present in a chromosome with the usual, localized kinetochore has been reported in several different species of plants by Kattermann (1939), Rhoades and Vilkomerson (1942), Prakken and Müntzing (1942), Östergren and Prakken (1946), and Vilkomerson (1950). Its occurrence poses some pertinent questions pertaining to the origin and nature of kinetochores in general. In this connection it is interesting to observe that all these cases and, indeed, all other departures from the more generally recognized localized kinetochore, have been encountered in but two groups of organisms—the monocotyledons among plants, and the arthropods among animals. Obviously no phylogenetic considerations can be based on this fact, but it may be fruitful to consider what it is in the structure of the chromosomes of these unrelated groups that lends itself to departures from the orthodox conception of the kinetic region.

The kinship between the kinetochore and the centriole has been pointed out by Darlington (1936) on theoretical, and by Schrader (1936) on both theoretical and structural, grounds. These are, for instance, their identical reactions to stains, their similarity in living cells, and their obvious functional relations in the mitotic cycle. The most striking piece of evidence in support of this relationship is the discovery of Pollister (1939) already mentioned. He has found that in the meiosis of the oligopyrene cells of certain viviparid molluscs, the kinetochores become detached from the degenerating chromosomes. These free kinetochores form a group with the centriole and duplicate its behavior and appearance exactly, in effect becoming extra centrioles.

*Aster*

It seems beyond question that the aster results from the activity of a localized region in the cytoplasm. The very extensive literature on the nature and origin of astral rays was given a judicious cytological survey by Wassermann (1929, p. 286), while the physiologist's viewpoint was presented by Heilbrunn (1928).

The fact that astral rays can often be distinguished so easily in the living cell is probably due mainly to the granules and other cytoplasmic elements that line their paths (Fig. 1, p. 7), although rays are sometimes also discernible in cytoplasm which is relatively clear. Chambers (1917, 1921b) considered that the rays are paths of centripetally moving fluid hyaloplasm. In his veiw, the centrosome represents this material gathering in the focal area. In the origin of the rays this more fluid material becomes separated from the granular constituents of the cytoplasm and it is these latter that constitute the gelatinous walls of what is essentially a system of radiating tubes or canals. Spek (1918), however, concluded that the physical condition of the centrosomal region is not always as fluid as the findings of Chambers would indicate. On the contrary, Spek believes that, as the aster develops, this region becomes more gelatinous.

A slightly different interpretation of astral rays was made by Pollister (1941), who agrees with the earlier workers that the astral rays are paths of flow. It appears probable from the researches of Bensley and Hoerr (1934), Mirsky (1936), Bensley (1938), and Banga and Szent-Györgyi (1940) that a considerable part of the basic ground substance of the cell (hyaloplasm) is composed of long molecules which Bensley calls "structure protein." Pollister suggests that in paths of flow or channels of diffusion these previously unoriented molecules become arranged in parallel formation, their long axes in conformity to the axis of the current. This would explain the birefringence observed in living asters by Schmidt (1936a, b, and c, 1937a, b), as well as the arrangement of long chondriosomes outside of the rays. The latter probably have been shunted out of the path of flow as the long molecules are arrayed in longitudinal formations. This in turn would explain why astral rays usually stain more heavily than the surrounding cytoplasm, a fact which has seemed paradoxical on the view that the rays represent watery channels. For if the rays are not only paths of diffusion but also parallel lines of oriented molecules, fixation might well tend to exaggerate this fiberlike arrangement and such a conformation would retain the dye more tenaciously than regions of unoriented molecules.

However, Inoué (1953), though not doubting that molecular or micellar reorientation is responsible for the optical appearance of astral rays, questions the existence of paths of flow which Chambers and Pollister hold responsible for such orientation. Inoué points out that in some material the rays end in a sharp point at the pole, a fact which is difficult to harmonize with a centripetal path of flow. Further, in Chaetopterus at least, Inoué can find no gathering of more fluid material in the centrosomal region, as Chambers claims for some other cases.

Finally it will be recalled that Heilbrunn (1920) suggested that the crenations in the membranes of centrifuged cells indicated that the aster is anchored rather firmly in the peripheral region through its rays or stands. A similar belief is implied by Conklin (1917) and Carlson (1952), and also constitutes a part of Boveri's hypothesis of mitosis (see p. 71). The difficulty of correlating such findings with the conception that the rays represent paths of more liquid cytoplasm is manifest. However on the hypothesis that rays also represent a reorientation of the molecular pattern so as to form fiberlike structures, no such difficulty is encountered.

All this evidence, together with the birefringence that has been observed in the aster by Schmidt, Inoué, and other workers, permits the conclusion that the astral rays represent an arrangement of long micelles in longitudinal, parallel rows. Whether such a micellar orientation is due to flow is another question. Such a flow would not in itself account for all the physical properties of the aster.

*Spindle Constituents*

It is well to keep in mind that the various spindle fibers of the metaphase configuration are located in a substance that is quite distinct from the cytoplasm. This substance, constituting the main mass of the spindle body, is probably derived, wholly or in greater part, from the extra-chromosomal contents of the prophase nucleus, the nuclear sap. Bleier (1930b) has put special emphasis on this material—his "Paragenoplastin"—as has Koerperich (1930), who called it "substance parachromosomique." Both authors consider it as a necessary medium of the chromosomes without which they cannot function and live.

Becker (1936 and 1938) and Wada (1935, 1941) have returned to the convictions of several early investigators, concluding that the spindle body is entirely intranuclear in origin. This does indeed seem to be the case in the plants investigated by them, and is also demonstrated most strikingly in cases of gonomery where the paternal and maternal nuclei, though lying in contact with each other, give rise each to its own spindle. To the classical examples of this latter peculiarity such as the Copepoda, Crepidula, Ciona, and Torpedo may be added the subsequently reported instances of the rotifer Asplanchna (Storch 1924) and Drosophila (Huettner 1924).

But it hardly seems warranted to extend these findings to the rank of a generalization. That cytoplasmic elements, at least in the form of fibers, may sometimes enter into spindle construction is quite evident from those cases where fibers arise between the centers in the cytoplasm long before the nuclear membrane has broken down. These are later incorporated in the metaphase spindle, which thus is compound in its origin (Fig. 6.)

The nature and origin of the three structural elements of the spindle—continuous fiber, chromosomal fiber, and interzonal connection—are considered in the following pages.

CONTINUOUS FIBERS. The continuous fiber, by arbitrary definition, is a fiber extending from one pole or center to the other in the ordinary mitotic figure. In fixed preparations its similarity to the astral ray has been recognized from the earliest days and intense study only confirms this first impression. A seeming gradation from astral to continuous fibers is present in many mitotic figures,

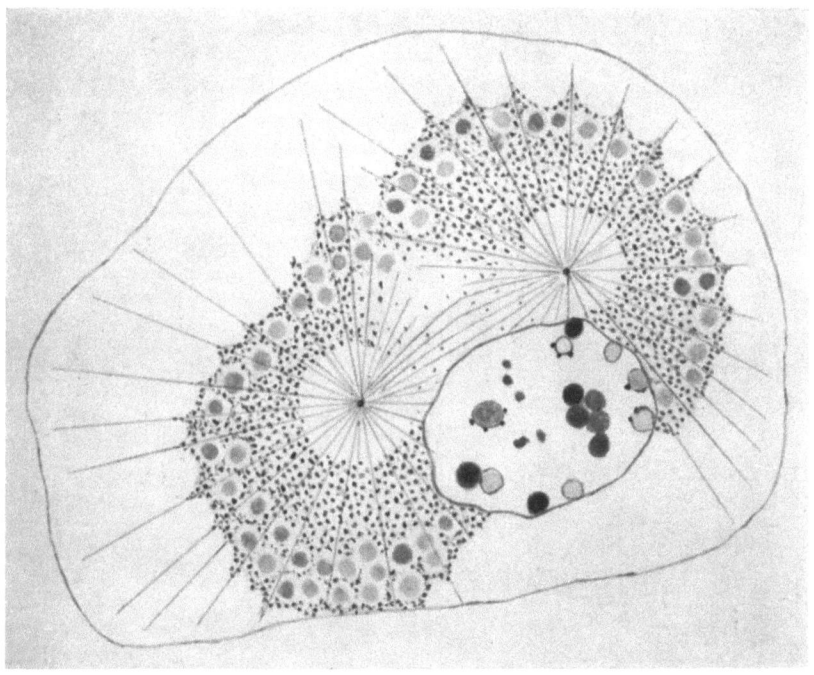

*Fig. 6.* Spindle formed in the cytoplasm before the nuclear wall has broken down. Second cleavage in the egg of the mollusc Arion (Lams 1910).

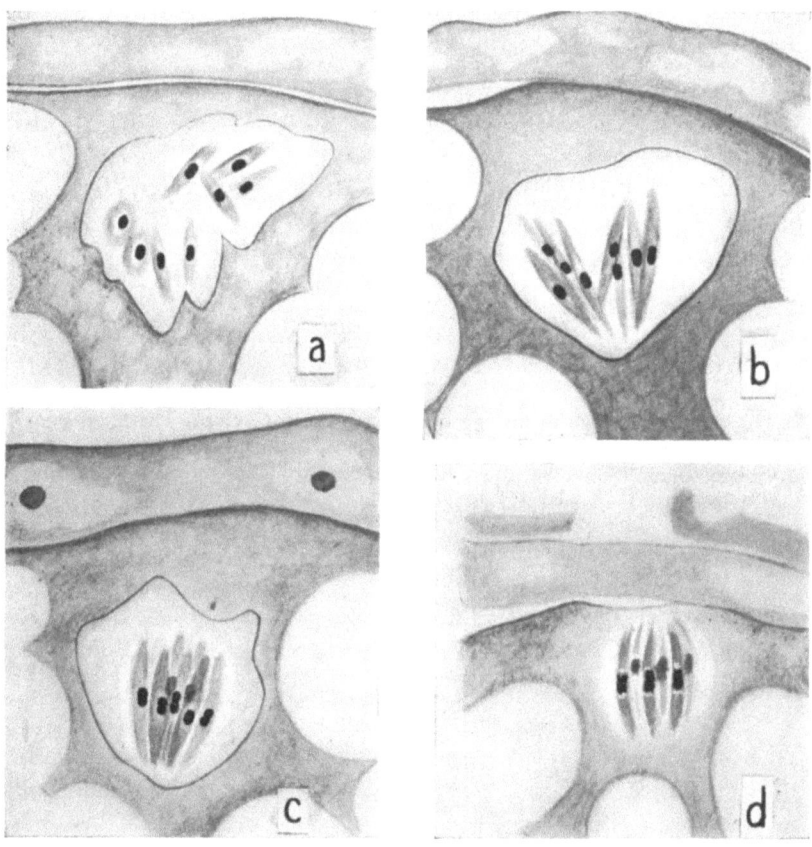

*Fig.* 7. Intranuclear origin of the spindle of the first meiotic division in the egg of the insect Acroschismus (Hughes-Schrader 1924). a. Prophase showing the association between chromosomes and chromosomal fibers. b. and c. Intermediate stages in the orientation of chromosomal fibers to form a bipolar spindle. d. Metaphase, shortly after the disappearance of the nuclear membrane.

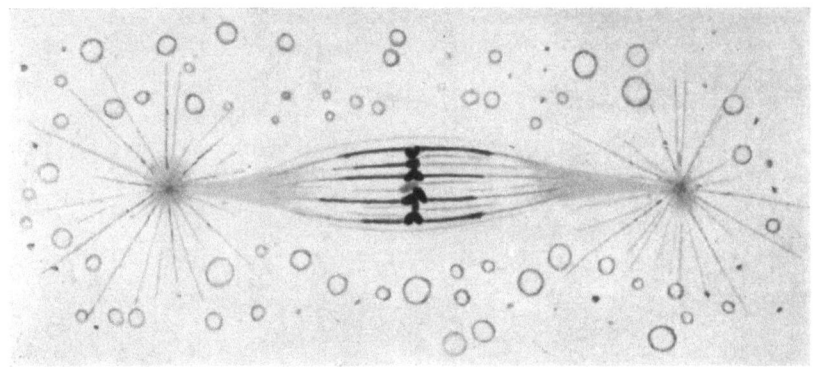

*Fig. 8.* First cleavage in the egg of the crustacean Cyclops. The fibers connecting the chromosomes with the poles show a thickening proximal to the chromosomes (Schrader 1934).

# STRUCTURE

as, for instance, in Fig. 6. It has been said that continuous fibers are frequently curved whereas astral rays are not, but even this is not always true, as witness the spiral asters of some molluscs (Painter 1916).

To view the continuous fiber as no more than a union of two astral rays that come from opposite centers is therefore in keeping with this observational evidence. So far as one can tell this should occur quite independently of the nucleus. The test might rest in artificially induced cytasters, which would be expected to show continuous fibers extending from one to the other. Rather often this does indeed seem to be the case, and the only difficulty lies in deciding whether we are dealing with an accidental, close approximation of two astral rays or a true continuous fiber. However, an extranuclear formation of such fibers is seen to occur in the normal course of events in several well-known cases. Among them are certain divisions in the salamander (Meves 1897a), as well as in the egg of the mollusc Arion (Lams 1910), where continuous fibers arise between two centers early in the mitotic cycle and form a miniature spindle (Fig. 6). This quite definitely occurs outside of the nucleus, and such a spindle gradually assumes metaphase proportions before the nuclear wall breaks down. Whether the nucleus has any influence on this spindle formation is not certain. Fankhauser (1934) believes that in certain anucleate cells of the salamander Triton a few continuous fibers may be formed between centers, but the figures indicate that it is difficult to be certain. At any rate, other cells that contained nuclei were closely associated with such cells and these could conceivably exert some influence on the process. Most decisive, therefore, would be cell divisions in embryos entirely devoid of nuclei. Despite Dalcq's doubts (1931) of successful divisions under such conditions, there now seems no question that they may occur (Harvey 1936, 1940). But in these cases, despite the presence of well-developed asters, no continuous fibers appear to be formed. In short, though continuous fibers can originate outside of the nucleus, the latter apparently has some influence in their formation.

But all this does not affect the claim that continuous fibers are, like astral rays, the result of polar activity and are essentially the same in structure. There is however one obvious difference between them as they are represented in the amphiaster at metaphase. In living cells the astral rays can be seen without difficulty, whereas continuous fibers are, with special exceptions, not visible. These exceptions (mentioned on p. 17) are certain mitoses in the mite Pediculopsis and several flagellate protozoa. In the former, Cooper (1941b) has observed continuous fibers in normally functioning cells. The peculiarity of this case lies in the fact that each chromosome constitutes a miniature nucleus or karyomere of its own, the wall of which disappears rather late in the mitosis. The beautiful mitotic configurations described by Cleveland (1934, 1935a, b, c, 1938a, b, c, d) in several symbiotic flagellates constitute the other exceptions. Both astral and continuous fibers can be seen in the living state with great clarity. The special conditions in these forms lie in the fact that the nucleus is maintained as such throughout the division (Fig. 18).[5] The chromosomes at metaphase are still with the confines of a nuclear membrane which persists through anaphase and, by constriction in the middle, also forms the walls of the two new telophase nuclei. In other words, the continuous fibers as well as the astral rays are extranuclear throughout.[5] It may well be that it is in these exceptional conditions that the visibility of fibers in the spindle is to be sought. If the Pediculopsis karyomeres retain their individuality even though their walls appear to break down, as Wada (1941) claims for orthodox nuclei, the fibers seen by Cooper are just as extranuclear as those of Cleveland's flagellates. This would strongly suggest that fibers are visible in the cytoplasm but become invisible as soon as they enter the nuclear substance—merely because of the optical properties of the two media. The test would be given by such cases as that of Arion. In the living cell the continuous fibers should there be visible before the nuclear membrane breaks down, but at metaphase when they are incorporated in the nuclear area they should fade from sight. A simple examination of this, or similar cases, should settle the issue.

[5] But Cleveland (1952) now is dubious about all his earlier observations on this point and deems it possible that some of the nuclear contents may get into the cytoplasm before metaphase. If that is the case, these fibers are visible in cytoplasm that has an admixture of nucleoplasm.

To sum up, the continuous fiber like the astral ray is a direct result of polar activity. The indications are that in basic structure the two are very much alike. Differences in their appearance and visibility are more probably a matter of the index of refraction and other physical properties of the various substrates involved than differences in the rays or fibers themselves. Perhaps this also accounts for the apparent absence of astral rays in most plants.

CHROMOSOMAL SPINDLE FIBERS. The chromosomal spindle fiber which is based on the kinetochore has naturally been regarded as the spindle element most intimately involved in chromosomal movement. Hence from the first, it has been the subject of special attention, and many hypotheses concerning its nature and origin have been broached, discarded, and advanced anew.

Without going into a historical review, it may be stated that basically these hypotheses simmer down to three: the chromosomal fibers arise from the pole, grow toward the chromosome and connect with it; or they arise from the chromosome and grow toward the pole; or they are formed as the result of an interaction between the pole and the chromosome. All this is entirely independent of the possible function of such chromosomal fibers.

The first of these hypotheses—that the pole originates the fibers and sends them to the chromosomes—is perhaps most firmly ensconced in the biological mind. It is also the oldest hypothesis and has most recently been sponsored again by Cleveland (1935c, 1938b). Indeed in many cases this would, superficially speaking, seem to be the most natural interpretation. That is especially true in many prometaphases where the fibers from the pole appear to exert pressure on the nuclear membrane, which becomes indented and crumpled as a result. However, consideration of all the features will show that this cannot be the whole story of the process. Even in the most striking of such cases the chromosomes cannot play an entirely passive role, for it must be remembered that the chromosomal fiber is not attached to the chromosome at random but solely and specifically at the kinetochore. Hence there can be no doubt that the latter has at least a directing influence. It must also be pointed out that the process of chromosomal fiber formation is exceedingly rapid, and so far as I know there has never been published a convincing series of stages showing a progressive growth of such fibers from the pole. Instead, all evidence indicates that after the initial stages in the disintegration of the nuclear membrane, the chromosomal fibers appear suddenly within the nuclear area and with their chromosomal connections fully established.

The view that the chromosomal fibers owe their formation to the chromosome, or rather the kinetochore, finds support in such cases as Acroschismus and many Coccidae (Hughes-Schrader 1924, 1942), where chromosomal fibers arise without reference to the location of the future poles of the spindles, and are oriented solely with respect to the individual chromosomes. Centralized polar forces only manifest themselves slightly later in bringing about a parallel arrangement of these chromosomal fibers (Fig. 7). Such a spindle therefore is clearly multiple in origin, its components arising primarily as a result of the activity of the individual chromosomes (Schrader 1932).

To sum up, the evidence indicates that the chromosomal spindle fiber may arise in two ways. It is the result of an interaction between the pole and the kinetochore in many cases, but in certain instances it may arise chiefly or entirely through the activity of the kinetochore alone.

The final configuration of the spindle may, however, be dependent on whether we are dealing with the direct or indirect type of spindle apparatus. In the former, the chromosomal fiber forms a direct connection between the pole and the kinetochore. In the indirect type, as conceived by Belar (1929a), the framework of the spindle is composed solely of continuous fibers and it is to them that the chromosomes become attached (Fig. 2). The latter is most strikingly shown in those cells where the spindle is formed in the cytoplasm outside of the nucleus and where the chromosomes become associated with it only after the nuclear membrane disintegrates. Fully to establish that such is the course of procedure is not easy, if only for the reason that a chromosome conceivably can form a direct connection with the pole after it has reached the spindle. Belar however was convinced that in Chorthippus he could see the substance that attaches the chromosome to the continuous fiber and conceived of it as a fluid secretion from the kinetochore which spread on the continuous fiber toward the pole. Wilson (1932) was strongly inclined to accept this interpretation and both Wada (1950) and Swann (1952) favor the concept of a secretion from the kinetochore on theoretical grounds. Such conditions seem to obtain also in certain cases other than Chorthippus, as, for instance, in the early cleavages of the crustaceans Cyclops (Schrader 1934), Artemia (Gross 1935), and Liogryllus (Kupka and Seelich 1948). (Fig. 8.)

Superficially it would seem that in this conception also there is an interaction between the kinetochores and the poles. But it will be recognized that at metaphase the two daughter chromosomes are still joined and a "fluid substance," if it follows the continuous fiber, must perforce flow in the direction of the poles.

The detailed structure of such a connection between kinetochore and continuous fiber is difficult to see. If it is homologous to the chromosomal fiber as might be expected, it should of course not be a secretion from the kinetochore but a reorientation of molecules. That does not exclude the possibility that in addition some substance is extruded by the kinetochore. In the chromosomal fibers of the direct type the structure is more open to analysis. Where large chromosomes are concerned, this structure is in the majority of cases a composite in the sense that the chromosomal fiber is made up of a number of very fine fibrillae (Fig. 9). But in any case, whatever a more detailed analysis may reveal concerning the connection of the chromosome to the spindle fiber, the basic structure of the latter is due to a reorientation of molecules or micelles.

Such a reorientation is involved also in the considerations of Freundlich (1927) who was the first to suggest that the whole spindle is a body which is characterized by an arrangement of its component particles that is analogous if not identical with the micellar orientation in a "tactoid" or liquid crystal. He rightly pointed out that if such is the nature of the spindle, a new light might be thrown on all the questions concerned with the problem of mitosis (p. 100).

A multipolar origin of spindles not directly associated with

42  STRUCTURE

chromosomes is difficult to interpret on almost any basis. It will be recalled that in certain plants (Osterhout 1897, and several other botanists) the late prophase was described as showing many small half-spindles. These were based on the nuclear membrane with their poles out in the cytoplasm. Only when the membrane broke

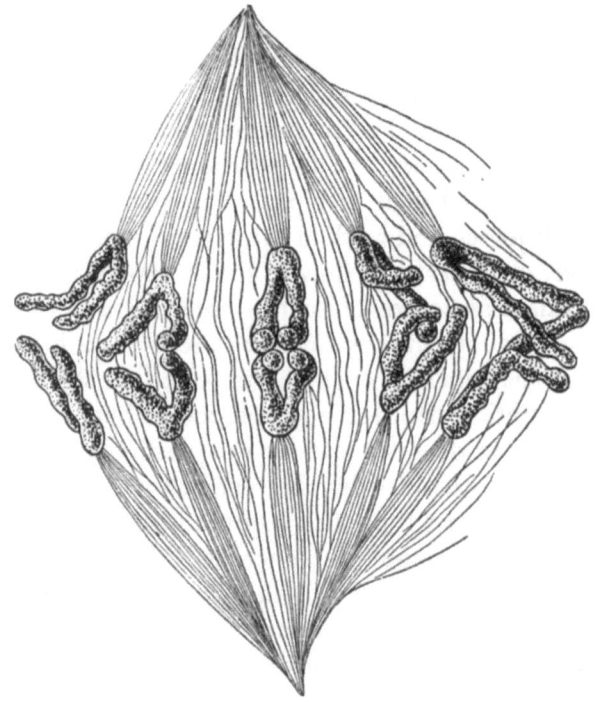

*Fig. 9.* First meiotic division in pollen mother cell of Lilium. Each chromosomal fiber is composed of a number of fine fibrillae. Continuous fibers are not connected with the chromosomes (Mottier 1903).

down did they become connected with the chromosomes and then they were bunched so as to form the orthodox, bipolar configuration. This description has been so generally accepted that the careful study of this question by Devisé (1922) did not receive the attention it merited. Devisé reported that the fibrous formations outside of the nucleus in prophase are actually chondriosomal in nature and that they take no part in the formation of the spindle. The latter is formed in the usual way with no multipolar phase at all. Similar conclusions were arrived at in several other plants by

## STRUCTURE

Jungers (1931, 1934), and it is possible that the supposed multipolar origin of plant spindles which finally become bipolar is based on misinterpretation.

The sum and substance of all this may be outlined in very few words. Without question there are important structural variations and differences in the types of fibrous elements that are associated with the spindle. But so far as their basic structure is concerned we are perhaps tilting at a strawman in trying to differentiate between astral rays and spindle fibers of all kinds. They all owe their character to a parallel orientation of long molecules. It is possible that there is a flow of cytoplasmic or nuclear substance in all these rays or fibers, but though there is some dubious evidence for this in astral rays (Chambers 1917, 1924), nothing comparable has yet been demonstrated for spindle fibers. So far as optical differences between the various elements are concerned, it is suggested that these rest primarily in the composition of the medium in which they are formed.

INTERZONAL STRUCTURES. The interzonal region comprises the nonfibrous ground substance of the spindle, possibly continuous fibers that persist in some form after metaphase, and, in many cases, interzonal connections that extend only between the ends of separating chromosomes.

If, as has long been believed, continuous fibers are present in the interzonal region they must have undergone some change after the metaphase that affects their physical properties. In many, though not all, cases, they are then no longer visible under the ordinary microscope even though they appear quite plainly in the half-spindle. In fact, as Schmidt (1936a, b, and c; 1937a and b; 1939) has pointed out for echinoderms, the absence of birefringence in the interzonal region indicates a loss of fibrous structure. Inoué (1953) agrees that there is indeed a marked loss of birefringence in the interzonal region but finds that in the worm Chaetopterus as well as in the lily, a weak manifestation of it is still present. He believes that the continuous fibers have not been broken down but have become hydrated and swollen, and this would account for the change in their optical properties.

But the continuous fibers of this region should not be confused with interzonal connections, wherever the latter occur—as, for instance, in the meiotic divisions of many Hemiptera. The difference between these two elements is especially clear in those forms where, as Wassermann (1929) and Bleier (1931) have pointed out, there are no continuous fibers visible even at metaphase. Nevertheless, such spindles often show the most striking interzonal connections, as can be demonstrated in many Coccidae. This is strong evidence for the claim that in such connections we are dealing

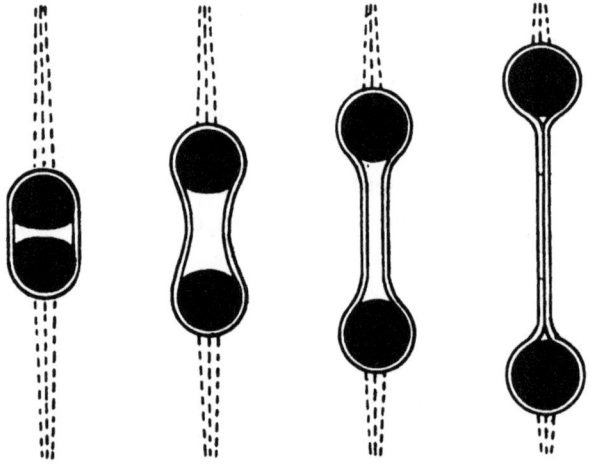

*Fig. 10.* Hypothetical origin of the interzonal connections from the pellicle. Progressive separation of the daughter chromosomes stretches the pellicle in which they are contained at metaphase.

with material not represented in metaphase fibers at all and possibly derived from the chromosomes themselves. Additional support for this is derived from the finding that chromosomes stained with methylene blue leave a streak of blue color behind them as they move to the poles (Jacobson and Webb 1951). The explanation that has been suggested (Fig. 10) is that these interzonal connections are nothing more than a sticky coating of the chromosome which is stretched like mucilage between the daughter chromosomes as they move further and further apart (Schrader 1932 and 1935, Becker 1933 and 1935, Upcott 1937a, and Carlson 1952). This coating is the pellicle or sheath which, unfortunately, is difficult to demonstrate in the metaphase chromosome although Hirschler (1942) has devised a technique that may obviate this difficulty. The tubular form of the interzonals in the coccids Llaveia (Hughes-Schrader 1931) and Protortonia (Schrader 1931) would support such a hypothesis, postulating merely that there the sheath is so viscous that the tube formed by the separating daughter chromosomes within it does not readily collapse. Perhaps we are dealing in the pellicle with a semipermeable membrane around the chromosome.

Ellenhorn (1933) has interpreted such interzonal tubes as nothing more than channels left in the spindle substance by the moving chromosomes. That such temporary channels are formed, especially where the medium is rather viscous, is of course quite possible. But it is not easy to see how channels would come to resemble the interzonal tubes of Llaveia, which appear to have definite walls. Likewise, the long persistence of interzonals for some time after division would suggest that they are more than mere trails.

Connections between separating anaphase chromosomes, when of this nature, are Feulgen-negative. There are, however, several types of interzonal connections which differ in their origin and are Feulgen-positive. Thus in some cases the chromatin or matrix becomes more liquid or sticky than is usual. As a result the separating chromosomes tend to be drawn out like taffy and form Feulgen-positive bridges in the interzonal region. This may occur normally, as in certain aphids (Ris 1942), or as an experimental effect—as after X-raying (Alberti and Politzer 1923, White 1937) or hybridization (Klingstedt 1939). Similar in appearance but somewhat different in causation are the cases in which there is difficulty in separating the terminal regions of chromosomes in division. Here some irregularity in the condition of the end or telomere may be involved (Schrader 1941b). A third type of Feulgen-positive bridge has its origin in an inversion or translocation of a section of the chromosome body. If, through such a structural alteration, two kinetochores come to be located in one chromosome and these go to opposite poles in division, the chromosome is stretched out between them to form a bridge (Darlington 1937, McClintock 1938). Such inversion bridges have been the subject of much study by

## STRUCTURE

cytogeneticists, many of whom seem to be unaware of the fact that not all chromatin bridges are necessarily of this type.

The difference between the interzonal and the half-spindle regions is evinced not only in their optical properties but also by their behavior under experimental conditions, as has already been mentioned on page 17. The half-spindle region with its continuous and chromosomal spindle fibers presents considerable resistance to distortion by centrifugation whereas the interzonal region, whether it includes interzonal connections or not, is bent and twisted rather readily (Schrader 1934, Shimamura 1940; see also illustrations of F. R. Lillie 1909, Morgan 1910, Andrews 1915). This would comport rather well with Inoué's suggestion that during anaphase the continuous fibers of the interzonal region become hydrated and swollen and thus offer less resistance than at metaphase. The interzonal connections derived from the coating of the chromosomes would of course have a minimal stiffness in any case.

But though the rigidity of the interzonal region is rather low, it may persist long after the time when the chromosomal fibers and indeed the entire half-spindle have disappeared. Moreover its fibrous structure frequently becomes obvious again in late anaphase and telophase, and this suggests that the hydration of continuous fibers which Inoué assumes for the anaphase is later reversed. This reappearing fibrous structure may persist even into the following one or two mitotic cycles. Hirschler (1935) studied these persistent, bunched interzonal structures (his "fusom") in a great range of cells and suggested that they may play a more important role in the life of such cells than has hitherto been suspected. It is of course possible that such a "fusom" includes interzonal connections as well as restored continuous fibers.

Some reference might be made here to the so-called "midbodies," those puzzling, deeply staining knobs or spheres that appear in the middle of the interzonal region at the telophase of many animals. They may be homologous to the much more extensive cell plate of plants, though this seems dubious in the light of recent researches (Conard 1939, Sinnott and Bloch 1941). Fry (1937) concluded that such midbodies are "nothing but the accumulations of dye at places where the spindle materials have been pinched together by the division furrow." But this interpretation seems insecure in the light of exceptions mentioned by Fry himself, where midbodies appear in the complete absence of a cleavage furrow. Nevertheless there is every likelihood that the formation of midbodies is closely involved with the behavior of the interzonal connections and continuous fibers. Possibly the latter are similarly involved in the formation of the phragmoplast of plants. To be sure, Van Regemorter (1926), Conard (1939), Sinnott and Bloch (1941), and others are convinced that the phragmoplast is a new structure arising in the entire absence of interzonal fibers. But, if Inoué (1953) is correct in his opinion that the absence of interzonal structures is only apparent, these opinions may have to be revised. At the same time it must be pointed out that in forms where continuous fibers fail to appear even in metaphase, their actual absence must be reckoned with.

To sum up, both continuous fibers and interzonal connections may be present in the ground substance of the interzonal region. The chief difficulty in a final anaylsis of this part of the spindle lies in determining the changes that the continuous fibers undergo while passing from the metaphase to the anaphase.

## CHEMISTRY

The chemistry of the cell constitutents is, as a matter of course, a fundamental feature in the analysis of mitosis. The discovery of the universal occurrence of nucleoproteins in cells by Miescher, Kossel, and their school during the seventies and eighties of the last century laid a sound foundation for the numerous biochemical investigations that followed. However, despite the great contribution of the biochemist, the limitations of biochemical methods when applied to some of our problems became more and more obvious. In addition to alterations in the native state of the cellular material that is introduced by extraction and fractionation techniques (such as partial loss, dissociation of compounds, and depolymerization), difficulties arise mainly from the fact that the biochemist is perforce dealing with cells en masse. Even with the most delicate methods (micromethod of Linderstroem-Lang; paper chromatography of Stein and Moore) relatively large quantities of cells must be used, and one is therefore dealing with an over-all picture. These unavoidable handicaps inherent in the biochemical techniques are particularly serious when one is dealing with the chemistry of the notably labile processes that are involved in mitosis. There the position as well as the interaction of the individual cell structures is highly important and calls for an analysis *in situ*.

The general recognition of these difficulties will explain why the new type of microscopy that was developed by Caspersson (1936) was immediately recognized as a large step forward in the chemical analysis of cellular structures. The outstanding feature in Caspersson's method lies in the fact that it makes possible a chemical analysis of individual cell structures *in situ*, directly under the microscope. Thus the morphology of the cell can be correlated directly with its chemical composition. The method involves the discovery that the absorption of ultraviolet light at specific wave lengths, by unstained cell structures, is due to certain chemical substances—notably nucleic acid and proteins. At present this type of cytochemistry is still restricted in its applicability because the number of substances recognizable in this way is not great, but fortunately the nucleoproteins are all-important building stones of the cell.

A second cytochemical method of analyzing the cell was more recently worked out by Pollister and Ris (1947). In this the cell structure in question is first stained with a specific dye. The amount of this dye is then measured by microspectrophotometry using the visible range of the spectrum. This method also has its limitations, for the number of dyes known to be specific for certain chemical substances and to follow the Beer-Lambert law is still small. However, refinements and new discoveries will probably broaden the range of both of these types of cytochemical analysis (Caspersson 1950, Pollister 1952) and further advances may be expected from using them jointly and in combination with biochemical techniques (C. Leuchtenberger, R. Leuchtenberger, R. Vendrely, and C. Vendrely 1952; Bryan 1952 in conjunction with Ogur, Erickson, Rosen, Sax, and Holden 1951).

As might be expected, the efforts of cytochemists to particularize our knowledge of cell chemistry to certain structures in the cell does not always lead to immediate and decisive conclusions. There is still much incomplete evidence and conjecture, and it is often difficult to separate them from definitely established findings.

The following brief summary concerns itself primarily with the chemistry of the nucleus. The final analysis of mitotic as well as all other cell activities will obviously involve the chemical interactions between nuclear and cytoplasmic components, but an appraisal of this field from the standpoint of chromosomal movements is at present premature.

During the past seventy-five years the chromosomes have been the most studied of all cell constituents. This is not only because of their interest for the geneticist but also because their ready stainability has made them conspicuous and rather easily followed constituents of the cell. For the cytologist the diagnostic and dominant component of the chromosome is what Flemming called "chromatin" ("the substance which is readily stained"), and nearly all biochemical and cytochemical findings unite in the conclusion that this substance constitutes the major part of the chromosome. It is a combination of desoxyribonucleic acid (DNA) with proteins, and it is the DNA that is primarily responsible for the more obvious staining properties of the chromosome (but see E. Stedman and E. Stedman 1947). Though it thus appears thoroughly established that the DNA is a prominent constituent of the chromosome, a final characterization of the proteins that are involved has not yet been made.

There is conjectural evidence that the synthesis of the chromosomal proteins is to be distinguished from the synthesis of extrachromosomal proteins (Caspersson 1947; Schrader and Leuchtenberger 1950), and evidence is accumulating that both basic and nonbasic proteins are present in the body of the chromosome. For instance, Mazia (1941) has suggested that basic proteins are associated with the skeleton thread or chromonema of the chromosome whereas nonbasic proteins constitute the main mass or matrix, but the final word on this subject is still to be said. Most of the information on the proteins in chromosomes comes from biochemical rather than cytochemical analyses. In these, two favorable materials have been utilized extensively. The first consists of chromosomes abstracted or isolated from the resting nuclei of certain tissues in considerable quantity (Claude and Potter 1943, Pollister and Mirsky 1943, and Mirsky and Ris 1947a and b). Whether such material really represents the actual chromosomes in their original condition or not—and this is still under dispute—it is probable that the major portion of the material involved is chromosomal in nature. The second type of material consists of the heads of sperms, chiefly fish sperms, which can be obtained in bulk (E. Stedman and E. Stedman 1943, 1947). Here too the chromosomal material constitutes much the largest part of the substance analyzed though it is known that small quantities of other cell structures are also present in sperm heads. The results of such analyses agree to the extent that in both materials there are found, beside the simple basic proteins of the histone or protamine type, more complex proteins which appear to be more acidic in character. While it is thus evident that several types of proteins are involved in the chromosomes, the results so far are still very incomplete and actually represent only a first step toward the systematic investigation that is called for.

There is also some evidence that small quantities of ribonucleic acid (RNA) are present in the chromosomes (see for instance Kaufmann, McDonald, and Gay 1951; Jacobson and Webb 1951). Thus the former belief that RNA is confined to the cytoplasm cannot be held any longer. Nor is DNA exclusively nuclear; in the cytoplasm of various eggs large quantities of DNA have been found (Mazia 1949; Zeuthen 1951; Elson and Chargaff 1952). It is an interesting question what its relation to the nuclear DNA may be.

But when all is said and done, we are still confronted with an acute lack of definite information concerning the chemistry of chromosomes, and the only well-established fact is that DNA is a constant and prominent constituent of all chromosomes. There is some danger that this may give to DNA a disproportionate importance in our considerations and cause us to lose sight of the possibility that other chromosomal substances which are less obvious may also play important roles. But that is a natural lack of balance in our outlook which will undoubtedly be corrected as we progress.

The nucleolus, which is present in the great majority of cells as a definite body or bodies, is composed largely of RNA and proteins. Caspersson (1950) concluded that the latter are predominantly basic proteins but the biochemical analysis of Vincent (1952), who succeeded in isolating large quantities of nucleoli from Asterias eggs, led him to conclude that the bulk of the nucleolus is phosphoprotein. Vincent, like Pollister and Ris (1947) before him, could find no histone in the nucleolus. In most organisms, though not all, it arises in close association with definite regions in certain chromosomes, but the nature of this association is by no means clear. In its activities, the nucleolus is correlated with the growth of both the nucleus and the cytoplasm, and the larger the nucleolus and its contained amount of ribonucleoproteins, the greater the amount of proteins in the nuclei and the cytoplasm (see the review of the extensive work of his school by Caspersson 1950; also Lagerstedt 1949, and Schrader and Leuchtenberger 1950).

The above information is still too restricted for our immediate purposes of mitotic analysis. However a few indirect conclusions can be drawn from it. Thus it appears that in Arvelius the size of the spindle (and of the centers) is larger in large cells with big nuclei than in cells with nuclei of smaller dimensions (Fig. 11). Since in these variously sized cells and nuclei the size of the chromosomes and amount of chromosomal substances (as measured by the DNA) are always the same, it would seem to follow that the dimensions of the spindle do not depend on the quantity of chromosomal substance but rather on the cellular size as well as the amount of certain proteins in the nucleus and cell. If the chromosomes contribute a part of their substance to the spindle, as is maintained by several investigators (p. 41), then such a chromosomal contribution is obviously not DNA. A method of gathering isolated spindles en masse, recently devised by Mazia and Dan (1952), promises to make possible a more detailed chemical analysis.

Another suggestion arises from a different aspect of cytochemical study. The division of a chromosome results in a halving of the chromosomal mass. Before the next division, the original mass is restored, and it is to be expected that the method and timing through which this restoration takes place will greatly affect the

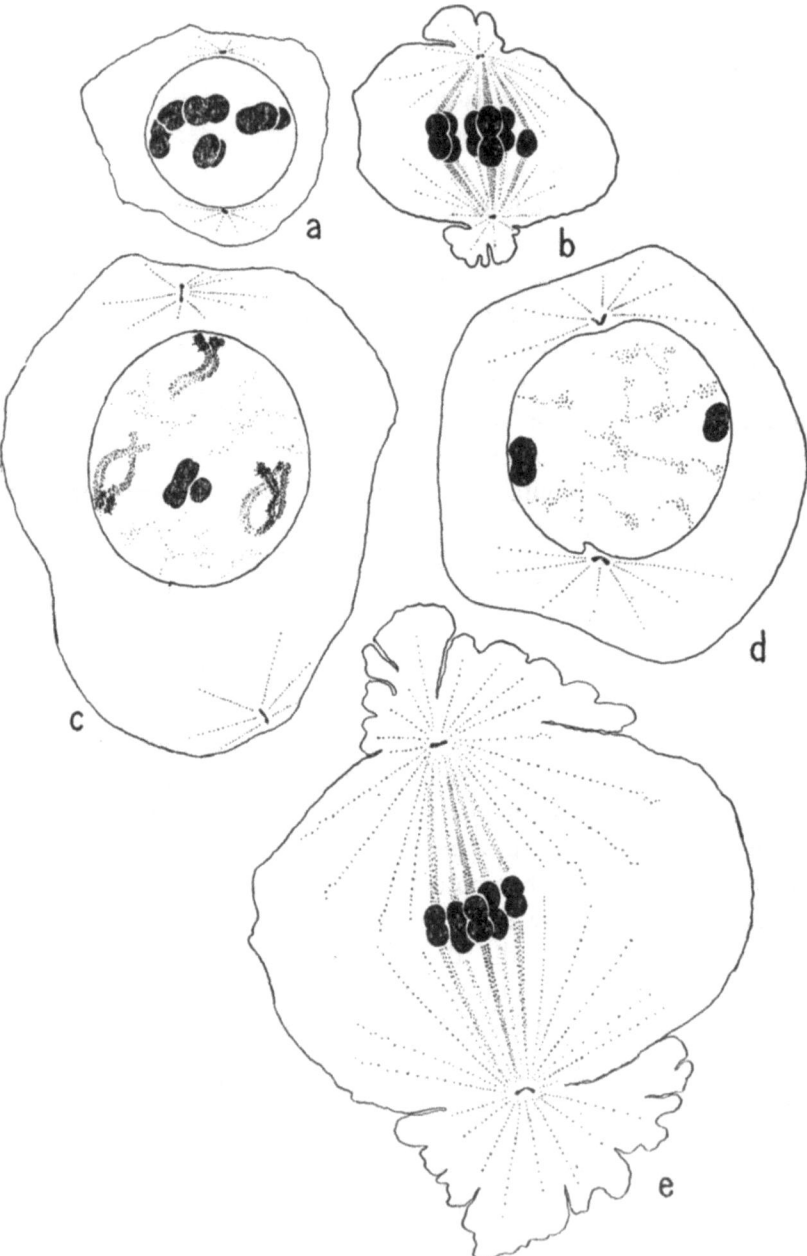

*Fig. 11.* Size of spindle in relation to other cell structures in Arvelius (Schrader 1947). a. and b. Diakinesis and metaphase in small cell with small nucleus. c., d., and e. Diakinesis and metaphase in large cell with large nucleus. Size of chromosomes is identical in all cells.

behavior of the chromosome. It is especially on the timing of this process of restoration that we are beginning to gather some useful information.

Pasteels and Lison (1950b, Lison and Pasteels 1950) have concluded that in various vertebrates and echinoderms the synthesis of DNA that is involved in bringing the chromosome mass back to normal occurs very rapidly during the telophase.

Swift (1950a and b), Pollister, Swift, and Alfert (1951), and Howard and Pelc (1951) found that this synthesis takes place during the interphase.

The results of Schrader and Leuchtenberger (1949), Leuchtenberger and Lund (1952), and Leuchtenberger, G. Klein, and E. Klein (1952) indicate that the synthesis of DNA does not always occur at the same time in different tissues though, generally speaking, it takes place not later than early prophase.

Despite these differences, the general conclusion is unmistakable; the synthesis of new DNA occurs soon after a mitotic division and does not extend through the entire prophase, as has so often been assumed. Furthermore, recent attempts of some cytologists to attribute certain chromosomal maneuvers to shifting and varying amounts of nucleic acid (see for instance the ruminations of Darlington 1947) would seem to be rather far removed from cytochemical evidence.

In all these considerations it is clear that as the biochemical and cytochemical researches progress, it is urgently necessary that contact with the cytologist be maintained. But such an obligation is reciprocal; cytology will make many false moves if it ties its findings to a chemical framework that is speculative.

For the present it must be realized that where more complex structures like the chromosomes are concerned, a mere determination of the chemical compounds which enter into their composition will not furnish a master key to chromosomal behavior in mitosis. A more exact localization of the various subtances in the chromosome body is needed before such knowledge can become really useful in our analysis. This means that further refinements of our methods are called for.

# III. Hypotheses of Mitosis

THE CHIEF CONCERN of those who have tried to elucidate the mechanics of chromosome movement has been with the phenomena of anaphase and early telophase. While these comprise perhaps the most striking and dramatic part of mitosis, it is manifest that the changes encountered in the prophase and metaphase are an integral part of the cycle and, indeed, that all the steps are interrelated. The mechanisms involved in the earlier stages are if anything more difficult of solution than the anaphase movement, as will appear in considering the various hypotheses that have been broached.

In evaluating these hypotheses, the two main problems that confront the investigator may be roughly put as follows: (1) How is the metaphase established? (2) What brings about the movement of the chromosomes to the poles? It is hardly necessary to state that each of these questions is accompanied by a complex of minor questions, and that these may turn out to be of paramount importance in the solution. It is indicative of our ignorance that we usually cannot see the bearing that the problems of prophase and interphase have on the points that puzzle us in metaphase and anaphase.

It may be said that the duration of the various steps that compose a complex biological action will often throw much light on the underlying mechanism. This is probably true of mitosis also, though there is considerable difficulty in deciding what constitutes the steps of the cycle and how they are delimited.

The duration of a complete orthodox mitotic cycle is extremely variable; it depends on the species, the age of the individual, the tissue involved, the temperature, and other factors. Indeed, Ephrussi (1926, 1933) and Bucciante (1927) found that each phase of the cycle has its own temperature coefficient. A great many measurements have been made (see Martens 1927 and Rabinowitz 1941 for references) and though the total time of a mitosis may vary from ten minutes to several hours, there seems to be general agreement about the relative duration of the different stages. The metaphase and anaphase are completed rather rapidly, whereas the interphase and prophase take a much longer time (Table 1). Strikingly divergent results are nearly always due to the fact that it is difficult to define and agree on the limits of telophase, interphase, and prophase.

TABLE I

DURATION OF STAGES IN THE MITOTIC CYCLE
(TIME IN MINUTES)

|  | *Drosophila Cleavage* (fixed cells) Rabinowitz 1941 | *Chick Mesenchyme* (live cells) Lewis and Lewis 1917 | *Arrhenatherum* (Plant) (live cells) Martens 1927 |
|---|---|---|---|
| Interphase | 2.9 | 30-120 | — |
| Prophase | 3.6 | 30- 60 | 36-45 |
| Metaphase | .5 | 2- 10 | 7-10 |
| Anaphase | 1.2 | 2- 3 | 15-20 |
| Telophase | .9 | 3- 12 | 20-35 |
| Total | 9.1 | 67-205 | 78-110 |

The absolute velocity of the anaphase movement of chromosomes is exceedingly variable for different types of cells and conditions. Using photographs for his measurements, Barber (1939) recorded this movement in staminal hairs of Tradescantia, chick fibroblasts, and the meiotic division of grasshoppers as varying from $.3\mu$ to $3.5\mu$ per minute. In other words the maximum was ten times as fast as the slowest rate. Though Barber's measurements were not exact, they serve to give some idea of the range of speed encompassed and comport rather well with the more accurate measurements reported later by Ris (1943).

In considering the mechanics of chromosome movement it is well to keep this absolute velocity in mind. The distance covered is at best extremely small, and it takes the chromosome a long time to traverse it. Even the comparatively high speed of 3 or $4\mu$ per minute mentioned for the grasshopper chromosome is many times slower than the movement of the hour hand on an ordinary watch. Most of us who have watched a motion picture of the mitotic process are inclined to forget that such photographic representations give a vastly exaggerated impression of both the velocity and the distances that are involved.

It must be recognized that the customary division of mitosis into stages has a rather arbitrary basis. Furthermore the timing of such stages constitutes a very crude beginning of an analysis, for any careful study of a given stage will quickly reveal that it is itself a complex of processes. It is the recognition and timing of the latter that will furnish really useful data (p. 79).

### EXPERIMENTAL ANALYSIS

Scientists in other fields are wont to express astonishment that the experimental method has been used with so little success in analyzing mitosis. To one who has tried it, the explanation is obvious. It lies in the fact that it is almost impossible to affect a given structure or process of the mitotic mechanism, be it by operation or physicochemical means, without simultaneously affecting several others. Usually it is difficult to determine the nature and precise influence of the disturbing factors thus introduced, and the intrusion of subsidiary effects may not actually become recognizable until a later period, if at all. This is true even of such exemplary experimental investigations as those of von Möllendorf and Wada, and these investigators have been duly cognizant of the complicating factors.

Despite these somewhat discouraging features, the last ten years have seen many new experimental attacks on the mitotic problem, and their total number greatly exceeds that of the first thirty-five years of the century. A survey of the far-flung experimental work that is now being pursued has recently been made by Hughes (1952) and, though not complete, it gives a good picture of the difficulties in the various methods of attack as well as the successes that have been achieved by them. In general, such work may be grouped under two headings: (1) A study of the methods that step up or decrease the division rate of cells in certain tissues or tissue cultures. (2) An analysis of treatments which recognizably affect certain cell structures or cell activities.

1. Acceleration and retardation of cell division has been brought about in a host of ways, ranging from temperature changes to

hormones and enzymes. As one scans this mass of work, especially that which is concerned with malignant growth, it seems hopeless to recognize a common denominator for all these agencies. It is patent that a multiplicity of problems is inherent in the question of division rate.

In the end, a narrowing of the attack will perhaps prove to be the most fruitful. Such a narrowing is involved, for instance, in the work of Lettré (see his review of 1948; also H. Lettré and and R. Lettré 1947) who for some time has made a systematic study of growth-promoting versus growth-inhibiting compounds, with division being considered indicative of growth. It is Lettré's endeavor to arrive at some basic principle involved in their action, but though his logically based investigations are making hopeful progress it is evident that a task of considerable magnitude still confronts him.

But such researches concern themselves with the cell as a whole. The answers, which they are beginning to furnish, do not have much immediate usefulness for the more specific questions of karyokinesis which concern us here, though they will no doubt become applicable as our knowledge increases.

2. An agency that affects only certain cell structures or activities would offer a better opportunity to gain some insight into the nature of mitosis. The ideal, of course, is a substance that is specific for a single, recognizable constituent of the cell. It is in its approach to this ideal that colchicine has assumed such great importance during the last fifteen years (by 1947 Eigsti had already listed some 950 titles in a bibliography of papers concerned with colchicine).

The general action of colchicine in the disorganization of spindle fibers and the inhibition of spindle action is now well known. But it is undeniable that in addition to this most striking effect on the spindle, colchicine also exerts an influence on the chromosomes (Östergren 1944b) and the cytoplasm (Beams and Evans 1940). Several other, seemingly unrelated, compounds also have an inhibitory action on the spindle, and one, podophyllin, is much more powerful in its effect than is colchicine (I. Cornman and M. E. Cornman 1951). But like colchicine, they too have subsidiary effects. In short, even the most favorable substances do not entirely meet the basic difficulty that attends the experimental attack on mitosis.

The most logical attack on the problem of mitotic poisons would seem to lie in a detailed analysis, one by one, of all the subsidiary effects that a single type of compound exerts on a certain type of cell. The beginning of some such task is represented in recent work of Zeuthen (1951) and though it involves much painstaking labor, the results should provide a sound basis for future advances.

At present it is difficult to estimate the forward steps that are being made by the experimental attack in the mitotic field. It appears that some time must elapse before a just appraisal can be made.

### THE PERIOD PRIOR TO METAPHASE

The interest of most workers in mitosis has been centered in the happenings that occur after metaphase has been established—that is, the anaphase. Indeed, the cellular phenomena prior to metaphase are so bewildering that this is quite understandable. Nevertheless it is not likely that with the metaphase an entirely new set of conditions is suddenly initiated. It is conceivable that the difference in the cytological manifestations between the two periods is in large part due to the presence or absence of a nuclear membrane and that the division line between pre- and post-metaphase is not as sharp as it seems to be.

In the following few pages an attempt will be made to outline the chief problems that arise in the analysis of the period before metaphase, together with the explanations that are at present available. For the sake of convenience, the prophase maneuvers are dealt with separately from the events that are more immediately involved in the establishment of the metaphase plate.

### PROPHASE MOVEMENTS

It has already been pointed out that the chromosomal maneuvers prior to anaphase present a kaleidoscopic roster of questions which often are so baffling that the problems of the anaphase movements seem simple by comparison. In giving a very brief discussion of

## HYPOTHESES OF MITOSIS

this difficult period in the mitotic cycle I am aware of the fact that the interpretations are usually little more than restatements of the descriptive accounts. Nevertheless they will serve to show where the chief difficulties lie, and that is not without value.

In considering the various questions involved in the prophase movements of chromosomes, it must be kept in mind that, in contrast to the period after metaphase, any interaction between elements within the nucleus and those in the cytoplasm must take place through the nuclear membrane. This, in itself, justifies drawing a dividing line between pre- and post-metaphase events and, to some degree, facilitates our analysis.

It is unfortunate that most of our knowledge concerning several of the prophase stages is based on studies of meiotic cells where there are special conditions which may not be expressed in other cells or are present in a more or less exaggerated form. The emphasis that has been put on meiotic cells by most workers is not merely due to the interest that they have for the geneticist, but results naturally from the fact that they offer technical advantages for cytological study. In any case, the forces underlying the mitotic aspects of meiosis must be present in other cells too and hence must be reckoned with.

There is some advantage in grouping the questions that arise from a consideration of these prophase events, since they are somewhat bewildering in their diversity. It will be realized that in view of our present ignorance of the underlying factors, such a grouping is rather arbitrary. The questions at issue are here taken up under the following headings:

1. Interaction between centers and chromosomes
2. Chromosomal maneuvers independent of centers
3. The establishment of bipolarity
4. Interaction between centers and non-chromosomal elements

1. The dominating influence on chromosome movements all through the prophase resides in the centers. Evidence for the close interrelation that existed between the centers and kinetochores during the preceding anaphase persists, in some cases, right through the following interphase and even into the prophase. This evidence

consists in the maintenance of the late anaphase placement of chromosomes, with the kinetochores gathered close to the center and directed toward it—the so-called Rabl orientation. However it is more than likely that once established, this orientation is a passive one and connotes merely that in some cases there is little shifting of the chromosomes.

But the Rabl orientation is not to be identified with the bouquet formation which is characteristic of the early prophase or leptotene stage in meiotic cells in so many species, and is perhaps reflected in the synezesis of plants. In both the Rabl orientation and the bouquet formation there is a definite orientation of the chromosomes toward the centers, but in the bouquet it is the ends of the chromosomes that are directed toward the centers and not the kinetochores (see discussion in Schrader 1941d, Hughes-Schrader 1943b). This means that though the center serves as the focal point of attraction in either case, the conditions in the chromosomes have undergone significant alterations. Incidentally, there is here further evidence for the claim that the terminal regions are specially constituted parts of the chromosome and that these ends or telomeres merit more attention than they have hitherto received (p. 116). In some species the first bouquet of the leptotene stage may lapse only to be succeeded by a second bouquet during pachytene (Hughes-Schrader 1943b), but that does not occur in the common course of events. It may be mentioned that the role of the center in the formation of the bouquet has been recognized for many years and is beautifully demonstrated in such papers as that of A. and K. E. Schreiner (1906).

An attraction between the chromosomes and the center may, in a few species, be evinced much later than the early prophase. One of the most striking cases is Anisolabis (Schrader 1941a), where the establishment of bipolarity is delayed until diakinesis. As the two centers separate and slide around the outside of the nucleus they are followed on the inside by the chromosomes which, as a consequence, separate into two groups (Fig. 14). Since the chromosomes are already in a late phase of condensation it is not possible to decide whether, here too, it is the chromosome ends that are attracted. Indeed, the fact that these cells are close to the metaphase stage may raise some doubt whether we are dealing with the same mechanism that underlies the bouquet formation. It should be remembered that in most species there is no attraction between centers and chromosomes at diakinesis; on the contrary, in cases such as Brachystethus and Mecistorhinus the autosomes tend to take a mid-point position, as far as possible from the two centers (Fig. 13). Since in both Mecistorhinus and Anisolabis the centers at this time are close to establishing the bipolar figure, it is not likely that there are decided functional differences between them. The cause for the difference in behavior must be sought in the chromosomes themselves, and this is borne out by the subsequent developments in Mecistorhinus (Schrader 1946b).

The evidence, diverse as it is, shows that there may be an attraction between the centers and the chromosomes, and that this hinges on the conditions that obtain in the chromosomes or certain regions in them.

2. Chromosomal maneuvers that are not directly influenced by the centers are presented especially in the synapsis period of meiosis. The pairing of homologous chromosome threads is, in most cases, a very precise and exact process of bringing corresponding regions of a pair of chromosomes into intimate contact with each other. The mechanics underlying this process are at present far beyond our knowledge (p. 115), but it may be pointed out that an attraction between homologous chromosomes may also occur in the non-meiotic cells of certain species (especially the Diptera) and that in such cells also the centers appear to play no direct role in the pairing.

Shortly before or during diakinesis occurs another movement of the chromosomes that has no apparent relation to the centers. In this, the chromosomes migrate to the nuclear periphery and assume a position in close juxtaposition to the membrane. Here they are more or less evenly spaced, and it is possible that the maneuver is due to surface tension conditions which are not manifested until the chromosomes reach a certain condition. The fact that the condition of the chromosomes plays an important role in this step is attested by the finding that the sex chromosomes, if heteropycnotic and hence differing in their state of coiling or condensation, may not be subject to such peripheral placement.

3. The movements of the centers can be followed with certainty only in the cases where visible centrioles or asters occur. One of their most striking maneuvers is their establishment of bipolarity. In effect, this comprises the separation of the two daughter centers (which arise from the single center that the cell has received in the preceding division) to opposite regions of the cell. This movement may occur in the cytoplasm without an obvious relation to the nucleus (see for instance Johnson 1931) or else it may take place in contact with the nuclear membrane on whose exterior surface the centers seem to slide in taking up a vis-à-vis position (as in Anisolabis). This variability in the interrelation of nucleus and centers must signify a corresponding range in their chemical and physical conditions. Nothing is known of the forces that bring about the separation of the centers. A simple repulsion obviously can not offer a complete explanation of the movement, though a hint that something like repulsion forces may at least participate is conveyed by the behavior of the centers in Anisolabis. In that insect there frequently are four centers (at a time later than the confused stage) and these usually come to rest in positions that are as far apart as it is possible to get within the compass of a sphere (Schrader 1941a).

4. It is likely that the differences in the behavior of the centers during the establishment of the future poles of the spindle are a reflection of varying conditions in the nucleus. In nearly all cases there is some period during the late prophase in which an attraction between the nucleus and the two centers is evinced. In the cases where the movements of the latter take place in intimate contact with the nuclear membrane, this attraction is manifest. In others however, the centers may first take up a position at opposite points of the cell wall at a considerable distance from the nucleus. Only in the succeeding phase is there cytological evidence of any interaction between them. If the centers retain their position at the cell wall, the rounded nucleus begins to elongate and may actually send out fingerlike processes to establish contact with them (Fig. 12). In other forms, the centers may desert the periphery of the cell and migrate to the nucleus (Fig. 11). It is pertinent to note

that if the attraction occurs in the confused stage, it does not appear to involve the chromosomes for, with rare exceptions, the latter undergo no shifts in position. This must mean that the

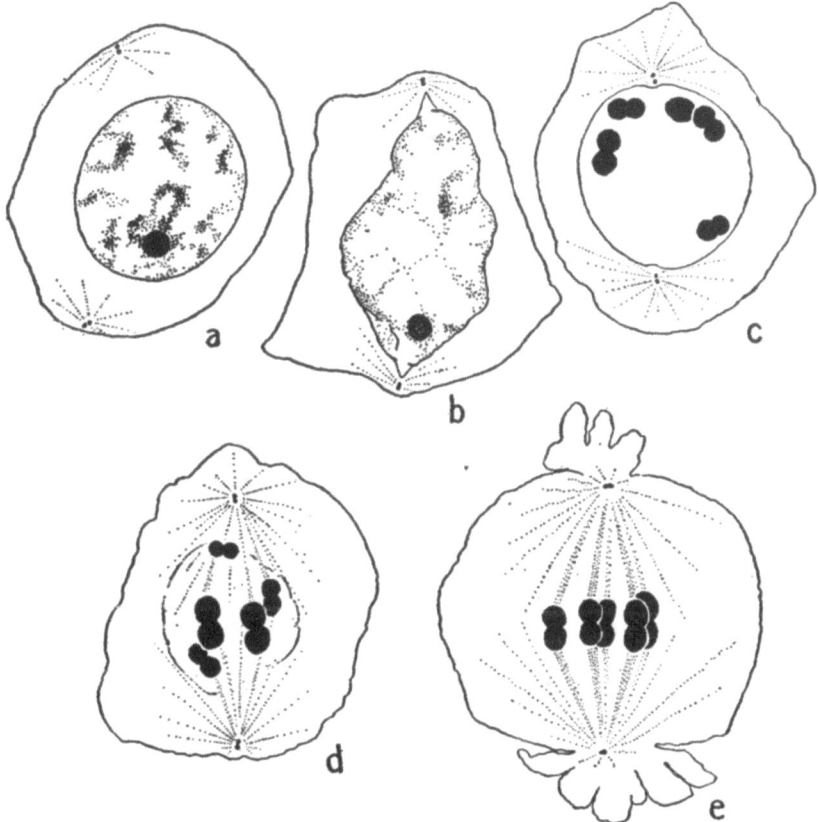

*Fig. 12.* Interaction between nucleus and centers in Loxa (Schrader 1947). a. Early confused stage with centers at cell periphery, not yet opposite each other. b. Middle confused stage; nucleus elongated in direction of centers at cell periphery. c. Late diakinesis; centers have left cell periphery. d. and e. Establishment of metaphase.

response to the centers rests in the nuclear sap or the nuclear membrane.

To sum up, though the altering conditions of the chromosomes greatly influence their behavior as the prophase advances, it is the center which serves as the influential pivot around which most of the maneuvers take place. But it is not amiss to reiterate that all these prophase maneuvers involve two facts which, though obvious, are sometimes forgotten by the analysts of the anaphase. They are:

1. Interaction between the prophase chromosomes and the centers must take place through or across the nuclear membrane.

2. The chromosomes have a wide range of movement which occurs without the intercession of spindle fibers.

### METAPHASE MECHANICS

The establishment of the metaphase plate presents a rather intimidating number of problems. None of the hypotheses dealing with this period in the mitotic cycle has fully surmounted all the difficulties involved. It is the variety of unknown factors confronting the investigator that complicates every attempt at an explanation. As a result, an appraisal of the different hypotheses when made within the confines of a book like the present one tends to be somewhat unfair. This is simply because it is not possible to give any characterization of these hypotheses that combines completeness with the necessary brevity.

A listing of the main questions may facilitate a better comprehension of what these difficulties are:

1. When, during prometaphase, the nuclear membrane disappears, the chromosomes are usually left scattered through the region that was formerly occupied by the prophase nucleus. How is their movement to an approximate mid-point between the two poles brought about?

2. The chromosomes finally come to be ranged in an exact, flat plane at right angles to the polar axis. What brings about this more precise movement?

3. In many metaphase plates, especially in meiosis, the placement of chromosomes is rather exact in the sense that certain chromosomes always occupy the same relative position in the metaphase configuration. What underlies such a distribution?

4. In the course of the prometaphase the chromosome, through its two kinetochores, becomes connected with each pole by a

# HYPOTHESES OF MITOSIS 65

chromosomal fiber. Why do not both chromosomal fibers frequently go to the same pole?

It will be objected that these questions do not cover all the special cases of metaphase behavior, but it is more than likely that those too will find an explanation if satisfactory answers can be given to the main questions listed above.

To begin with, it may be stated as a general rule that no regular metaphase plate is established unless a bipolar spindle apparatus is present. That is conclusive evidence that here too the two poles or centers exert a deciding influence.

1. The first movement of the chromosomes toward the general mid-region has been explained on various grounds. A propulsion due to electric forces, viscosity changes, currents, etc., is considered under the hypotheses of anaphase movement taken up in later pages. At this time only the more recent explanations (which are not necessarily new) will be discussed. They are covered by the terms "pulling" and "pushing."

The basic conception of a pushing mechanism is an old one and was taken up again by Belar (1927, 1929a and b). In his interpretation, the continuous fibers are formed when fibers originating at the two opposite poles meet and join in the mid-region. In growing out from the poles, these fibers push the scattered chromosomes before them toward the middle. Wada (1949a and b, 1950) made use of the same general idea but did not commit himself to continuous fibers as such. He conceives of the fibrous structure of the spindle as arising from the folded protein chains in the polar regions or caps (he is dealing with plants). In unfolding, these chains push the chromosomes before them into the mid-region. Wada (1950) supplements this pushing action with "forces caused by the bipolarity" through which a chromosome is moved toward the equator "just as anything in the field of gravity—falls down toward the center of the earth."

2. The placement of the chromosomes in a flat plate is considered by Belar to constitute a separate step. According to him each chromosome becomes attached to a continuous fiber by a viscid fluid secreted from the kinetochore. This fluid spreads out over the continuous fiber to some extent and is called "traction

fiber" by Belar. It allows a to-and-fro sliding motion on the continuous fiber, and, utilizing this freedom of motion, the chromosome finally comes to rest in the exact equatorial plane.

Although Wada also argues for a secretory fluid from the kinetochore he ascribes no role to it in metaphase formation. Evidently he considers the mechanism proposed by him for the first step adequate to bring about the second as well. Östergren (1945b and c, 1949a, b, and c, 1950a, 1951), who more than anyone else has concerned himself with the problems of metaphase mechanics, also considers that one and the same process is involved in the first two steps. In his explanation he returns to the hypothesis of Drüner (1895), which was also adopted by Meves (1897b) and has been considered more recently by Wassermann (1929), Rashevsky (1941), and Schrader (1944, 1947a). Each chromosome is conceived to be subjected to a pull in the poleward direction, and, as Östergren emphasizes in his hypothesis, the force of this pull increases with the distance between the kinetochore and the pole. Since such a pull is also exerted toward the opposite pole, the chromosome finally comes to rest midway between, in a position of equilibrium.

It may be said that some sort of give-and-take action during prometaphase is evident when one observes the to-and-fro movement of chromosomes in living cells, and it is dramatically shown in the internationally known film of K. Michel. The nature of the pull in Östergren's interpretation, which is considered on page 74, does not affect this aspect of the hypothesis.

Östergren rightly points out that such a hypothesis is more concrete in its formulation than one that postulates an interpolar region in which ill-defined forces help to gather the chromosomes —as is implied in both Belar's and Wada's explanations. Further, since his hypothesis calls for nothing but traction forces, it can also lay claim to greater simplicity. However in certain respects he is guilty of oversimplification. It appears that under some conditions, the chromosomal fibers may push as well as pull, and in such cases as Mecistorhinus and Brachystethus (Schrader 1946a and b, 1947a), where a whole group of chromosomes is shifted forcefully toward the side of the cell, the marked elongation of the chromosomal fibers strongly suggests a pushing action. Hughes (1952) attributes this sideward displacement of chromosomes to the transverse force of Östergren, but it seems highly improbable that such a force would ever be strong enough to cause the distortion of the cell that is frequently shown (Fig. 13). Also, in considering Östergren's hypothesis, it must not be forgotten that in some instances a chromosome may gain an equatorial position

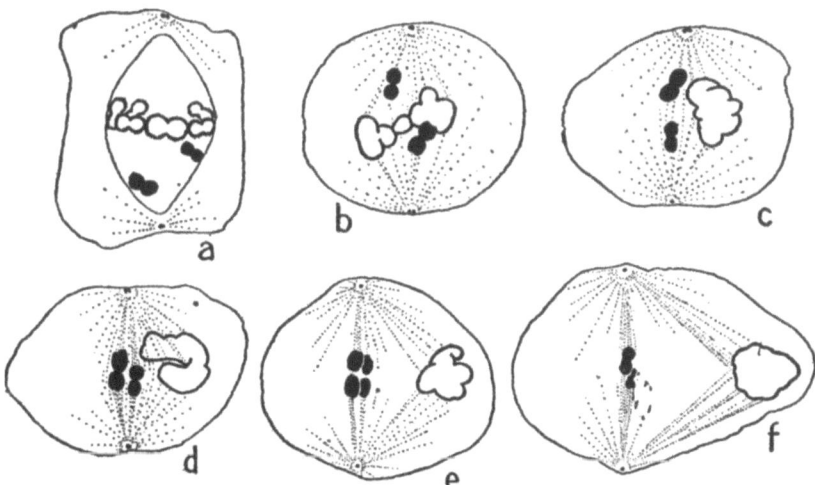

*Fig. 13.* Lateral displacement of autosomes in Brachystethus (Schrader 1946a). a. Late diakinesis; autosomes midway between centers. b., c., d., and e. Establishment of metaphase with increasing displacement of autosomal clump. f. Extreme displacement in Mecistorhinus (Schrader 1946b). Sex chromosomes drawn black in all figures.

even though there is cytological evidence for a chromosomal fiber to only one pole (Hughes-Schrader 1948b). Finally, forces not involving fibers of any kind are certainly active in the chromosome movements prior to the disappearance of the nuclear membrane (p. 59). Such observations indicate that, though the supposition of unknown forces gathering the chromosomes in the mid-region may come close to begging the question, the existence of factors other than those considered by Östergren must nevertheless be reckoned with (Schrader 1947a).

3. The placement or distribution of chromosomes with respect

to each other, in the metaphase plate, has always been difficult to explain. It is in this connection that the conception of the spindle as a system of mobile particles in a dynamic equilibrium has proved to be especially useful (Östergren 1945b, 1949a). Any body like a chromosome would locally alter the arrangement of these particles, and this would increase the potential energy of the system, giving rise to forces which tend to shunt the chromosomes out of the spindle in a generally sidewise or transverse direction. But since the chromosome is kept in the general interpolar region by traction forces from the two poles, its final position is the resultant of these poleward and transverse forces.

Östergren stresses the fact that since the ejecting force would be stronger for large than for small chromosomes, the latter would tend to take a more central location in the plate while the larger ones would be at the periphery. In actuality the larger chromosomes are by no means always placed peripherally and the reasonable assumption would have to be made that the two forces involved may vary under different conditions.

Further if, as Östergren believes, we are dealing with a tactoid, nucleoli and chromatoid bodies should be shunted out of the spindle with greater freedom than the chromosomes, since they are not subject to poleward traction forces. Actually they seem to undergo no transverse shifts at all.

Withal, the hypothesis of Östergren meets with fewer objections than any other, especially if we think of the spindle in the more general terms of a dynamic system rather than specifically a tactoid.

The fact that the metaphase chromosomes do not come in contact with each other has been given various explanations. R. S. Lillie (1905b) postulated a mutal repulsion which followed logically from his assumption of electric charges on the chromosomes. Metz (1934) and Östergren (1943) sponsored the possibility that the swollen chromosomes of interphase gradually shrink so that the spaces between them at metaphase represent nothing more than shrinkage spaces filled with nuclear sap. Possibly too, we are dealing with surface tension phenomena similar to those which determine the spacing in diakinesis or even earlier (Schrader

1947a). In any case, this problem of spacing does not seem to pose insurmountable difficulties to a final explanation, though discriminatory tests are called for.

4. Interaction of the chromosomes with the poles, and the establishment of chromosomal fibers, are primarily functions of the kinetochore. Since at metaphase every chromosome or bivalent has, in effect, two kinetochores, it becomes a question how their activity is regulated so that the two chromosomal fibers go to opposite poles instead of to one and the same pole. The problem has received very little attention in the past and only Östergren (1950a, 1951) has dealt with it *in extenso*.

Östergren's primary postulate is that the kinetochore in the metaphase and anaphase chromosome is not radially symmetrical but has an akinetic as well as a kinetic side. When two kinetochores are present, they become oriented "back to back," with the kinetic sides facing outward in opposite directions. Interaction with the center or pole can only occur if the kinetochore faces in that direction, and in its prometaphase movements it will probably do so, by chance, sooner or later. The pull on the kinetochore that then automatically ensues when interrelation with the pole is established forces its companion kinetochore to face in the general direction of the opposite pole, and thus the bipolar orientation is completed. We may for the present ignore the minor difficulty that arises when chromosomes are so placed as to make connection with the second pole difficult or impossible, as occasionally happens in Homarus and Loxa (Schrader 1947a).

The greatest difficulty for Östergren's attempt to explain the orientation of kinetochores arises from cases where two or more chromosomes form a linear chain, as in the $X^1X^2Y$ of mantids. Here it is difficult to see why a pull on one of the three kinetochores toward one pole will automatically face the kinetochore of an adjacent chromosome toward the opposite pole, for at this phase the chromosomes of some species are long and obviously flexible. The same difficulty arises in the consideration of the alternate orientation of chromosomes in the long chains of Oenothera. At best, no such simple mechanism will suffice to explain instances like those of *Perlodes microcephala* (Matthey and Aubert 1947), where the three components of a compound X frequently form a chain that always passes to one pole in the reduction division. Indeed, we are dealing here with a puzzle of long standing which is even greater when the components of such a compound X show no material connection with each other and nevertheless always pass to the same pole, as in the case of several nematodes (Walton 1924).

Östergren frankly considers that several aspects of his reasoning are more or less speculative. Like all other interpretations of prometaphase maneuvers, it is forced to leave the exact identification of the forces involved to the future. Thus the placement or orientation of two kinetochores back to back involves a "mutual interaction between the kinetochores" that we know nothing about. It is likely that we are here dealing with the larger question of chromosomal orientation in general—a question that has already arisen in connection with the pairing of chromosomes, and induced Cooper (1938) to postulate a special pairing surface in each chromosome, and Piza (1943) to speak of dorsoventrality.

In summation, it is evident from the foregoing discussion that the establishment of the metaphase is more distant from a final explanation than almost any other part of the mitotic cycle. Östergren's attempt to furnish a basis for further exploration is highly to be commended and his working hypothesis is at present the most useful available.

#### POST-METAPHASE MOVEMENTS

Most investigators of mitosis are primarily concerned with the mechanism involved in the anaphase movements of the chromosomes. This will be apparent from the following pages which deal with what are ordinarily conceived of as mitotic hypotheses. With few exceptions these consider the events prior to metaphase only superficially, and such exceptions have been given special treatment in the pages immediately preceding this one.

#### CONTRACTION: PULLING

In 1878 and 1879 Klein published observations which led him

## HYPOTHESES OF MITOSIS

to conclude that certain fibrous elements in the cell are contractile. However, his conception of the amphiastral configuration with which he was dealing was only an approximation of its true structure. The first definite formulation of a mitotic mechanism involving chromosomal spindle fibers capable of contraction and elongation came from van Beneden (1883).

This hypothesis, perhaps the most natural and appealing, may be outlined in a sentence or two: The metaphase chromosomes are in some way attached to the chromosomal fibers which in turn are based or anchored in the more or less stationary pole or center. By a simple contraction of these fibers the chromosomes are pulled to the pole.

A mitotic mechanism in such a simple form had a strong appeal and for many years was the one most generally accepted in textbooks. Among cytologists, however, it soon lost ground and came to be looked upon as entirely too naive and ill-comporting with a more exact knowledge of the mitotic apparatus. To mention just a few of the more specific objections, there was the finding that in many cases the chromosomal fibers end at an appreciable distance from the centriole (when one is present) whereas the anaphase movement often takes the chromosomes beyond this point. Less frequently the chromosomes were reported to move even beyond the centriole itself. Again the bulk of evidence indicated that the chromosomal fiber does not increase in diameter between metaphase and telophase whereas a contracting, elastic spindle fiber—usually conceived of as being analogous to a rubber band—would thicken as it shortened.

This last-named objection was, superficially at least, overcome by Boveri (1888) in his hypothesis, according to which the contraction of the chromosomal fibers plays a minor role, and most of the movement is due to a separation of the poles from each other. In such a movement the chromosomal fibers serve mainly as rather inactive connections between centers and chromosomes, the latter simply being dragged after the center. The centers themselves are moved as a result of the contraction of astral rays that in turn are anchored in the periphery of the cell. But this explanation merely shifts the difficulty, for although Boveri thought he could discern a progressive thickening of the astral rays, this seemed hardly striking enough to be convincing.

Shiwago and Troukhatchewa (1940) suggested that a pulling action on the part of the chromosomal fiber might be brought about, without recourse to the disputed elastic action, by assuming a progressive decrease in the volume of the fiber which itself is primarily an outpouring from the kinetochore. Such a hypothesis would of course account for any failure of chromosomal fibers to thicken progressively in anaphase.

Regardless of the optical difficulties in deciding many of the points at issue, the various objections gradually brought the traction hypothesis into disrepute. It was only during more recent years that the arguments in its favor were once more considered seriously (see Cornman 1944; Schrader 1944). This renewed interest is perhaps a natural consequence of the more recent conceptions of spindle structure and a recognition of the fact that the older analogy between traction fibers and rubber bands was too crude to be of much use beyond furnishing a first suggestion. Indeed it was perhaps chiefly this adherence to a macroscopic concept of the action of chromosomal fibers that laid the traction hypothesis open to attack.

While some workers (for instance Fujii and Yasui 1936, according to Shimamura 1940; and Rashevsky 1941) continued to espouse an elastic action of chromosomal fibers without entering into the details of the process involved, those who were concerned with the underlying mechanism began to seek an explanation of the whole anaphase movement in the physico-chemical basis of fibrous action. In the main the resulting interpretations can be given under two headings.

The first of these has profited considerably from the analysis of muscular action which in recent years has been especially the work of Szent-Györgyi (see his 1951 general account). The original interpretation of findings on contracting and expanding fibers was proposed by Meyer in 1929 (see also Meyer and Mark 1932) and somewhat later by Astbury (1935). According to this the myofibrils are composed of chains of protein molecules, and contraction results from a folding of such chains. A more direct application of these findings to the spindle fiber was made by

Schmidt (1937b, 1939) through his researches on the birefringence of mitotic spindles. He noted that the marked birefringence of metaphase, when the chromosomal fibers are greatly extended, gradually decreases during anaphase as these fibers shorten. His conclusion was that this is due to an action similar to that observed in myofibrils which also lose birefringence as they shorten or contract—in brief, a folding of protein chains. Swann (1952), who has also observed this loss of birefringence in the anaphase spindle, reports that in echinoderms the birefringence always disappears first in the immediate neighborhood of the kinetochore and thence progresses poleward. This suggested to him that it is the kinetochore that is primarily responsible for this alteration in the spindle, perhaps through some substance that it exudes, and that therefore it is the kinetochore that shortens the chromosomal fiber. However Inoué (1953) was unable to see in the cells of Lilium that the initial loss of birefringence began in the proximity of the kinetochore; in fact, birefringence seemed to be stronger in that region than in the rest of the half-spindle during early anaphase.

There are several objections to the protein chain hypothesis. Thus Östergren (1949a), like Piza (1943) before him, noted that prometaphase chromosomes may slide over each other without any apparent hindrance from the attached chromosomal fibers. This is a movement that is rather difficult to conceive of if protein chains constitute such fibers and give to them the character of more or less resistant cords. To escape this difficulty, Schrader (1951) suggested that at this early stage, prior to metaphase, the protein chain organization may still be incomplete and thus permit the passage of other cell elements through them. Inoué proposed the possibility that such protein chains are easily broken but may be reformed more rapidly than is generally thought.

It is unfortunately true that our knowledge of protein chains is still rather fragmentary and that, therefore, arguments such as the foregoing, pro and con, rest on an insecure basis. Thus an unfolding or denaturation of protein chains is usually irreversible in experiments performed in vitro though, as Goldacre (1952) pointed out, the living cell must have means of folding and unfolding its protein chains all the time.

A second objection voiced by Östergren presents itself in the fact that during prometaphase, when the chromosomal fibers are formed, some chromosomes are closer to one pole than the other. Since such chromosomes finally attain an equatorial position in which protein chains to either pole should have identical degrees of folding, it must be assumed that initially the chains to one pole are laid down in a partially folded condition whereas those to the other are extended. This laying down of protein chains with a seeming foreview of the fact that in the final equatorial position the protein chains to both poles are in the same state of extension is rather difficult to explain in the present state of our knowledge.

Objections such as these induced Östergren (1945b, 1949a) to abandon the protein chain hypothesis, while Wada (1949a, 1950) utilized it only in explaining the stage prior to anaphase. Both investigators consider the spindle to be a system of particles in a dynamic equilibrium in which such particles preserve parallel orientation despite shifts in position. The chromosomal fibers are regarded as special regions in which there is a closer packing of the oriented particles due to the influence of the kinetochores. This of course conforms to the general conception of the chromosomal fiber that other modern workers hold, but whereas Östergren believes this packing to be due to distance effects of kinetochoric forces, Wada attributes at least a subsidiary influence to a secretion from the kinetochore that holds or cements the particles of the chromosomal fibers together.

The transport of the chromosome to the pole is then held to be due to the strong tendency of the specially massed particles in the chromosomal fiber to return to the more stable distribution of particles in the rest of this dynamic system (the spindle). If this return to stability occurs especially at the poles, there would be a shortening of the chromosomal fiber which in turn might draw the chromosome in that direction. But this introduces another difficulty, namely a mode of attachment of the chromosomal fiber to the chromosome that would make it possible to exert a pull on the latter. Wada is inclined to seek some explanation of this in a kinetochoric secretion which finally connects with the pole, while Östergren conceives of the anaphase movement as due to a pull of the nature of surface tension.

Both Wada and Östergren are free to admit the speculative and tentative character of their hypotheses, which, like the protein chain hypotheses, must await further elucidation of the physicochemical aspects of the various factors involved. But as a result of investigations such as these, the traction hypothesis has once more assumed a very important position in our analysis of mitosis. Not only has it furthered our understanding of the cytological background but it also has helped us in approaching a stage where a more precise formulation of the necessary physicochemical attack is possible.

### EXPANSION: PUSHING

A conception that, in a sense, is just opposite to that of a pulling or traction mechanism, was first proposed by Watase (1891), based chiefly on his study of cleavage in the egg of the squid Loligo. According to this, fibers growing out from the two opposite centers exert pressure on the nucleus situated midway between them. The nucleus thus becomes flattened and its chromosomes are forced into the equatorial plane. When the nuclear membrane breaks down, the spindle fibers extend into the nuclear region and some of them become attached to the chromosomes. The growth of the latter fibers then continues and those from one pole gradually push certain chromosomes to the opposite pole and vice versa. The essential point is that here is given a mechanism postulated on an expansion of chromosomal fibers which is continuous from prophase to telophase and involves no complicating reversals of action.

There are several mechanical difficulties in this hypothesis. One of the greatest lies in the attachment of the fibers to the chromosomes so as to bring about an orderly division. In a somatic metaphase, the halves of a chromosome like that of Loligo have not yet separated, although they are oriented so as to put the plane of the "split" between them at right angles to the spindle axis. This is only saying that each chromosome-half is then closest to the pole toward which it will travel at anaphase. If it is to be pushed by a fiber from the opposite pole, this fiber must pass and avoid the chromosome-half nearest to it in order to reach the one which it has to push.

This, like several other difficulties, was not realized when the hypothesis was broached simply because the structural relationships of the chromosomes to the spindle fibers were then not generally known. One could make subsidiary assumptions to save the hypothesis, but these would run so contrary to observed facts that the effort seems doomed to defeat. Watase's conception of mitosis might well be abandoned.

### VARIATIONS: CONTRACTION AND EXPANSION

Several of the more recent hypotheses combine pulling and pushing processes in explaining the anaphase movement of chromosomes. Pulling is generally attributed to the contraction of chromosomal fibers whereas the pushing action is sought in some other constituent of the spindle, be it the interzonal region, the spindle as a whole, or structural elements in either.

The mitotic hypothesis of Belar (1927, 1929a, 1929b) already mentioned in connection with metaphase mechanics (p. 65) was a leading influence in resuscitating interest in mitotic problems, and served as a starting point for many other investigations. Its main points may be given briefly as follows:

1. The basic framework of the spindle is constituted of the continuous fibers which stretch from one pole to the other.

2. Each chromosome becomes attached to a continuous fiber by a viscid fluid secreted from the kinetochore, and this fluid follows or spreads out on its continuous fiber to the nearest pole. This fluid constitutes Belar's traction fiber.

3. The initial separation of the daughter chromosomes is an autonomous process.

4. The part of the continuous fiber located between the separating daughter chromosomes, as well as the nonfibrous substance of that region, begin to expand. They act as a pushing body or "Stemmkörper."

5. Each traction fiber glides on the continuous fiber (which serves as a guide or track) toward the pole. It may also shorten somewhat. The gliding motion is comparable to the transportation "of a food particle on the axopodium of an Actinosphaerium."

As already mentioned, this combination of traction and longitudinal expansion of the interzonal region bring about the anaphase movement.

Belar pointed out (1929a) that his hypothesis represented only a slight alteration of the older hypotheses of Drüner, Meves, and Bonnevie. The new feature lies in Belar's conception of the traction fiber. This, as already stated, he regarded not as a fiber coming from the pole but as originally a viscous fluid secreted at the kinetochore. Belar's researches on the subject were productive of many findings which stand on their own merit, regardless of the fate of his "working hypothesis." The weakness of the latter lies in the role ascribed to the "Stemmkörper," as well as the nature of his traction fiber. The share taken by the "Stemmkörper" was finally regarded as quite secondary by Belar himself as intimated in the last paper dealing with his work (Belar and Huth 1933). However, since it has played such an important role in older hypotheses, and since it may still be considered as the point that initiated Belar's researches, its weakness may well be pointed out here.

In the first place there need be no doubt that the interzonal region is subject to elongation in a great many species. It is, however, more than questionable whether expansion is confined to the interzonal part of the spindle. Thus Belar (1929a) himself induced a considerable elongation of half-spindles at metaphase, when there is as yet no "Stemmkörper" (by immersion in hypertonic media), and Ris (1942) confirmed this in aphids. Another demonstration that elongation is not necessarily correlated with the separation of daughter chromosomes, i.e., with the presence of an interzonal region, is furnished by some cases in Anisolabis in which an irregularity of the centers results in a failure to establish bipolarity. The chromosomes then do not establish a metaphase plate and undergo no division. Though there is consequently a failure to form an interzonal region, there is nevertheless a great expansion of the spindle which may cause sufficient pressure to distort the cell walls (Fig. 14). An equally great elongation of the spindle was encountered by F. Smith (1935) in the meiosis of the plant Impatiens. Perfectly normal tetrads are formed but, as in Anisolabis, no metaphase is established and, before the tetrads divide, the spindle has become extremely long. The great

78  HYPOTHESES OF MITOSIS

expansion of the spindle in certain hybrids (Brieger 1934, Dobzhansky 1934) often involves a "Stemmkörper," but the drawings indicate that the rest of the spindle may also participate. In short, the elongation in all these cases either does not involve the inter-

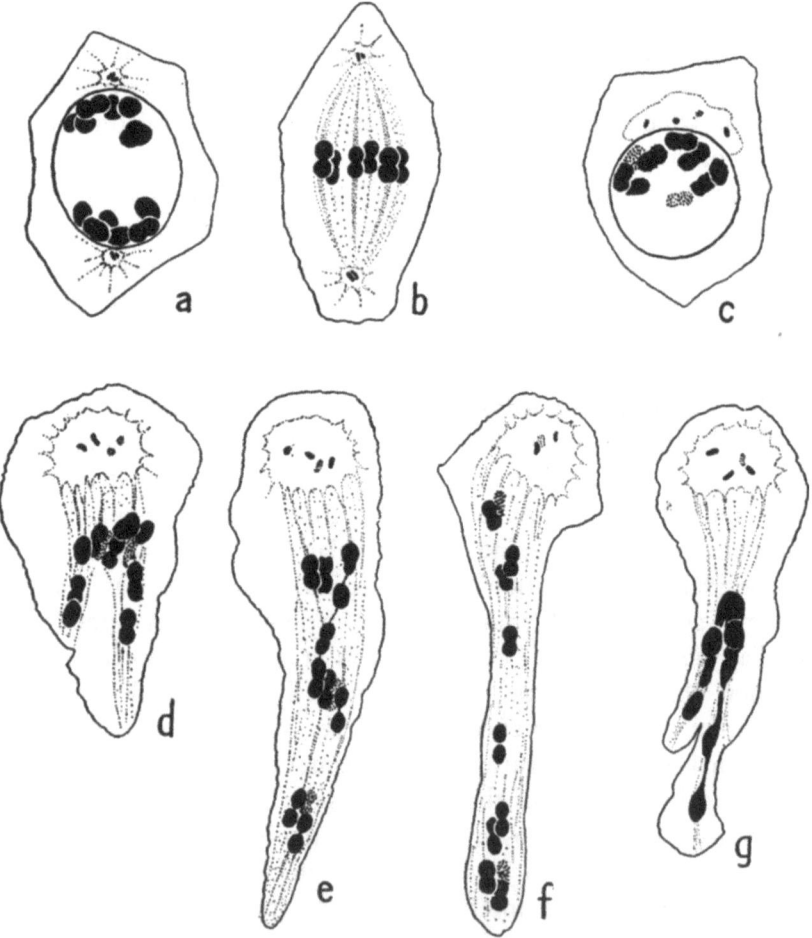

*Fig. 14.* Spindle fibers in bipolar and monopolar figures of Anisolabis (Schrader 1947a). a. Normal late diakinesis with chromosomes grouped close to centers. b. Normal metaphase with two centrioles at each pole. c. Late diakinesis in abnormal cell with all four centrioles together. d., e., and f. Monopolar figures showing elongation of spindle fibers without chromosomal division. g. End of monopolar division; undivided chromosomes clumping.

zonal region at all or else is due to the activity of other spindle regions as well.

A more general objection to the idea of a "Stemmkörper" lies in the fact that in many species the chromosomes do not all move at the same rate. Indeed, in certain hybrids there is not only a lagging of the univalent chromosomes but sometimes even a contrary movement during anaphase. As Bleier (1939) points out, this is hardly possible if there is an expansion of the interzonal region as a unit. This same difficulty would in fact be encountered by any hypothesis which assumes an absolute uniformity in the movement of a group of chromosomes.

This brings us to a consideration of the role played by Belar's traction fiber. It has been said that the existence of a fluid that flows out over the continuous fiber from the kinetochore is difficult of proof and Ris (1949) seriously doubts whether this indirect type of connection between the chromosome and the pole (Fig. 2b) ever occurs. That it is not altogether devoid of supporting evidence has already been pointed out (p. 40). Nor is Belar's vagueness in explaining the forces that bring about a sliding of the traction fiber necessarily fatal to his hypothesis, for such a sliding might well be attributed to a phenomenon like surface tension. But there can be no question that various aspects of Belar's hypothesis should be bolstered by more evidence before it can receive very serious consideration.

No such difficulties are encountered by the hypotheses that regard the traction or chromosomal fiber as a structural element whose action is relatively independent of the rest of the spindle. The careful observations and measurements of Ris (1942, 1943, 1949) on mitosis in the living cells of various insects lead to the conclusion that the anaphase movement consists of two separate processes or steps. In the grasshopper the two processes overlap to a considerable extent, but in certain Hemiptera most cells show the anaphase movement as taking place in two steps. In the first of these the chromosomal fibers shorten, i.e., the chromosomes approach the poles. In the second, the whole spindle elongates, with the result that the daughter groups of chromosomes are separated still further than they were by the first movement. In cells observed in chick tissue cultures by Hughes and Swann (1948), there is likewise a combination of these processes which in that case occur almost simultaneously.

The expansion or elongation of the spindle is often very great. In cases such as Anisolabis (Schrader 1941a) the distance between the poles at telophase may be double that of the preceding metaphase, and this is not an extreme instance. This elongation may have a pushing force of considerable magnitude as can be adduced from the fact that the cell wall may be bulged out under its pressure while the group of chromosomes, propelled before it, is pushed to one side (see, for instance, the case of Llaveia described by Hughes-Schrader 1931).

But even though this pushing factor thus plays a role in a great many if not most anaphases, it is not an absolutely necessary part of mitosis. In the first place, though such a pushing action undoubtedly aids in separating the daughter groups of chromosomes from each other, it does not bring the chromosomes closer to the poles because the latter are moved simultaneously in the same direction. In other words, a contraction of the chromosomal fibers or shortening of the half-spindle is absolutely necessary to reach the poles. Indeed, in some cases such as the crustacean Artemia (Gross 1935) there is no spindle elongation at all and the entire anaphase action rests in a contraction of the chromosomal fibers. Again, in cells of the grasshopper Chortophaga, Ris (1949) succeeded in halting the elongation of the spindle body with chloral hydrate without hindering the progress of the chromosomes to the poles.

There has been very little consideration of the physicochemical aspects of expansion. In general, investigators like Ris, and Hughes and Swann, think of the spindle as a whole without regard to the constituent elements such as continuous fibers and interzonal connections. The fact that no continuous fibers can be seen in certain cases (p. 44) which nevertheless show an elongation of the spindle would seem to cast doubt on their importance in the process. But Inoué's findings with the polarization microscope recommend an attitude of reserve wherever the absence of any spindle structure is concluded from evidence obtained with the ordinary microscope.

The importance of the role played by forces of traction has

already been evaluated in a previous section (p. 70). The present considerations show that expansion must also be considered, and that in most cases the two go hand in hand. Our cytological knowledge of the structures in which such expansion is based is by no means complete, but recent years have seen progress in that direction. We are now much closer than we were ten years ago to being able to ask the physical chemist specific questions.

### VISCOSITY AND HYDRATION

A mechanism based on an orderly sequence of viscosity changes in the cell has been propounded above all by Wassermann (1929). This hypothesis is central to nearly all of the reasoning concerning mitosis as presented in his extensive monograph, and finds expression for instance on page 339 (freely translated): When we see the formation of a spindle—in other words a progressive gelation advancing from the poles to the equator—occurring simultaneously with a propulsion of the chromosomes into the metaphase plate, then we may interpret this correlation to mean that the change in viscosity is equivalent to movement, or better, that it brings about movement.

In its general conception this hypothesis is striking in its simplicity. An increase in viscosity starting at the two poles, advances, toward the equator. The chromosomes distributed more or less at random in the middle region after the breakdown of the nuclear membrane are pushed thereby into the equatorial plane. The anaphase movement is due to a reversal in the direction of the process. After the slight, initial, autonomous separation of the daughter chromosomes, the interzonal substance is in turn subject to an increase in viscosity. As this progresses from the equator to the periphery of the cell, the two groups of chromosomes are pushed farther apart and finally to the poles. The hypothesis thus presented in its skeleton outline does not do justice to Wassermann's detailed considerations of the various complications that may be encountered and only the key idea is given here. A serious objection to such an hypothesis has been voiced on various occasions by Bleier (1930a, 1931, 1933, 1934, 1939). It is based chiefly on the meiosis of certain plant hybrids in which the bivalents divide normally whereas the univalents lag or even migrate in the opposite direction toward the equator. Very rightly, Bleier argues that such diverse movements occurring simultaneously in different chromosomes are not consistent with any hypothesis in which the means of propulsion advances as a wave, which would of course affect all chromosomes alike.

This objection is strikingly valid in the first spermatocyte of Sciara where certain members of a group of chromosomes migrate toward one side of the cell while the rest move a short distance toward the single pole. The movement starts from a group of chromosomes in which the homologues are not paired nor is there any metaphase plate. Metz (1933, 1936) therefore suggested that the selective nature of this segregation must originate in the activity of the individual chromosome. In conformity with Wassermann's hypothesis Metz (1933) argues that "this activity operates by bringing about a progressive alteration in the physical state (i.e., by solation or gelation) of the protoplasm adjacent to the chromosome." Metz further discusses whether this mechanism is centered in the sheath or the body of the chromosome, but that is not essential to the point at issue here (p. 106).

It may be stated at once that any hypothesis based on changes in viscosity alone can offer no valid explanation for chromosomal movement. The term viscosity is applied to the resistance of a fluid to deforming forces. Alterations in this resistance cannot result in pressure on a passive body and hence the latter will not be moved by such changes.

However, if the proponents of such hypotheses have in mind the expansion or swelling of certain regions of the cell, then they are standing on a more secure basis. In such an interpretation the expansion is nearly always the consequence of hydration and would be quite capable of bringing about a movement of the chromosomes before it. This indeed partially represents Wassermann's more recent viewpoint as given in 1939, although even there he does not draw the distinction very sharply.

All this is not saying that viscosity may not play a role in mitosis. Thus it might be expected that a more viscous spindle substance is correlated with a slower anaphase movement than a more liquid one. Barber (1939), however, concluded that the viscosity does not affect the velocity of the chromosomes, since long ones seem to move just as fast as shorter ones that should offer less resistance. However the absolute velocity is at best so low that such resistance may play only a minimal part, and Barber's conclusion that "Stemmkörper" action is indicated by the equal velocity of all chromosomes is not necessarily valid.

Conard (1939), on the basis of his study of the rather specialized conditions in the alga Degagnya, believes that the metaphase is established by a contraction of the nuclear volume due to dehydration. According to him the poleward movement of the chromosomes then results from an immense expansion by hydration of the substance in the narrow space or split that is visible between daughter chromosomes at metaphase. In other words, Conard substitutes hydration for the viscosity changes of Wassermann in explaining the anaphase movement.

The agency of swelling and contraction of certain regions in the cell has also been given the central point of importance in von Möllendorf's experimental studies of mitosis (1937 to 1939). It is clear that hydrating and dehydrating agents affect the various phases of mitosis very differently, and local inhibition and transfer of water may well play an important role in the general course of mitosis. Von Möllendorf is, however, properly cautious with respect to the more detailed aspects of chromosome movement and supports only the general view that, for instance, in the course of the anaphase "a process of swelling plays an important role in the transport of chromosomes toward the centers" (1938a, p. 51).

Incidentally it will be recognized that a swelling or expansion of a certain, circumscribed region is also represented in Belar's "Stemmkörper" action. The objections encompassed by Bleier's observations, however, confront all such hypotheses, as has already been noted.

One can only agree with von Möllendorf that the expansion and contraction of localized regions in the cell may well be of the utmost importance in cell mechanics. But in such general form the conclusion has no immediate application to our more specific problems. Indeed the relation of such changes to the complicated spindle apparatus is not at all clear, though there can be no doubt that the latter constitutes a necessary basis for the orderly movements of the chromosomes after prophase.

# 84  HYPOTHESES OF MITOSIS

## ELECTROSTATICS

The resemblance of an amphiaster to the configuration assumed by iron filings in a polarized magnetic field was emphasized as early as 1873 by Fol, and since that time a great many workers have attempted to explain chromosomal mechanics on the basis of magnetic or electrical forces. The difficulties encountered in such attempts are reflected in the extensive considerations of Gallardo, Hartog and Hardy, and have been summarized by Wilson (1925).

The most comprehensive hypothesis was that of R. S. Lillie (see especially 1905a and b and 1911), which is in many ways similar to that suggested somewhat later by Bernstein (1912). Lillie's hypothesis was put forward with due cognizance of the fact that it could be no more than a tentative beginning in an analysis of a highly complex mechanism. The basic idea was that chromosomes under certain conditions behave like electrified particles and that they move in an electrostatic field.

Unlike most other workers, Lillie tried to explain the origin of the two poles or centers. He suggests that both cell and nucleus have semipermeable membranes that are polarized as shown in Fig. 15a, with the cytoplasm electronegative and the nuclear contents—exclusive of the chromosomes—positive. The centers or poles originate through a local increase in ionic permeability of the cell membrane. In these two localities a free interchange of $+$ ions and $-$ ions becomes possible. The resulting depolarization progresses rather slowly and as a consequence the underlying regions of cytoplasm remain negative for some time. Colloid particles oriented along lines of force between them and the depolarized periphery represent the astral rays (Fig. 15b).

Simultaneously, the nuclear membrane also becomes depolarized. Again as in the case of the cytoplasm, the inner region of the nucleus is last to be affected and thus remains electropositive for some time (Fig. 15b). The chromosomes, which are electronegative, remain grouped in the electropositive middle region. But, being repelled strongly by the two negative centers, they become arranged in a plane between them. This is the metaphase plate where the chromosomes remain separated from each other because they of course mutually repel.

## HYPOTHESES OF MITOSIS

Each chromosome at this time is already internally divided. The daughter chromosomes now repel each other toward the two centers. This is the anaphase movement, and it presented what seemed to be a great difficulty since these chromosomes would have to move against the repulsion of the powerful centers (Fig. 15c).

Although this was at the time considered a weighty objection, it would seem on the premises of the hypothesis rather easy to overcome. As the depolarization of the two negative regions in the cytoplasm (the centers) progresses, their powers of repulsion

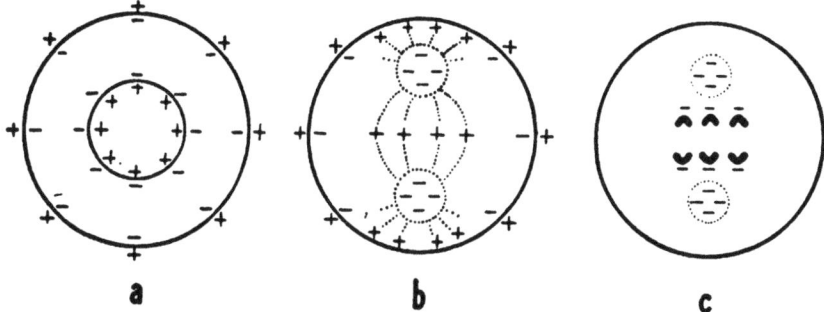

*Fig. 15.* Diagrams to show R. S. Lillie's electrostatic hypothesis. a. The cell before metaphase. Cellular and nuclear membranes are polarized. Cytoplasm is negative and nucleoplasm is positive. b. The establishment of two poles and a spindle through an increase in the permeability of the cell membrane in two opposite regions. The depolarization of the nuclear membrane occurs simultaneously. The positive region in the middle represents the nucleoplasm. The negative chromosomes are not shown. c. The conditions at anaphase, when the negative chromosomes move toward negative poles.

would of course wane and hence the chromosomes could complete their poleward migration. Such a series of events presupposes rather exact timing of the various processes involved, but exact coordination is a *sine qua non* in any hypothesis covering such a complex mechanism as mitosis.

Lillie's hypothesis can hardly be maintained in its original form at the present time and this is particularly true of his ideas concerning the origin of the poles. Concerning chromosomal movement it is almost certain that the series of events actually involved represents an almost kaleidoscopic panorama of alterations and conversions. Far from being too complicated as was once held, Lillie's hypothesis is an oversimplification. Nevertheless it may be pointed out that its basic reasoning has become part and parcel of many of the theoretical considerations promulgated by cytogeneticists during recent years, and is often used in other fields also.

Hypotheses very similar to that of Lillie have been advanced more recently by several investigators. Thus may be mentioned that of Kuwada (1929), who assumes, however, that the negative chromosomes become positive just before anaphase and therefore are attracted by the negative centers. This important change he considers to be the result of contact of the primarily negative chromosomes with the electropositive cytoplasm at the time of the disintegration of the nuclear membrane. Zirkle (1928), who in brief outline proposes a similar hypothesis, suggests that such an alteration results from the entrance of electropositive nucleolar substance into the chromosomes. The charge on the nucleolus of course explains why it migrates to the pole if it persists into anaphase. The advantage of both Kuwada's and Zirkle's hypotheses is to obviate the difficulty of the approach of the chromosomes to the centers in late anaphase. Both investigators assume the two centers to be negative but unlike Lillie do not consider the origin of such negative charges. But in a slightly later investigation of the root cells of Pinus, Zirkle (1931) concluded that the electrostatic role ascribed to the nucleolus in his earlier paper could not be upheld. However the claim that some nucleolar substance is contributed to the chromosomes is by no means new and has recently been advanced again by Jacobson and Webb (1951). If this can be fully established, it will constitute a factor that will have to be considered in every analysis of chromosomal movement, quite independently of its importance in any electrostatic hypothesis.

It may be remarked here that in his careful analysis Kuwada for the first time in a comprehensive hypothesis of mitosis introduced the kinetochores as important agents in the mechanics of chromosome movement.

Kuwada and his students were especially interested in the positions taken by chromosomes in the metaphase plate. Like Lillie (1905a) and Cannon (1923) before them, they experimented with floating magnetized needles subjected to the superior force of a larger magnet suspended above the basin. They found that such

## HYPOTHESES OF MITOSIS

needles can be induced to enter into arrangements that simulate the placement of chromosomes at metaphase. This would indeed support a rather simple conception of an equilibrium of repulsions and attractions which may nevertheless give a hint of actual conditions. However one must not lose sight of the fact that this is reasoning from an analogy and hence not without danger.

The hypotheses of Koller (1934), Upcott (1936), and Darlington (1936a, 1937) approach that of Lillie even more closely. Indeed they differ from it chiefly in making use of the kinetochore, which was practically unknown when Lillie proposed his hypothesis. Darlington furthermore builds into the hypothesis his assumptions concerning the meiotic behavior of chromosomes, which leads him to consider prophase phenomena as well as those later stages that are under attack in other hypotheses. Both Koller and Upcott have accepted Darlington's interpretations and many, perhaps a majority of cytogeneticists, analyze their findings on that basis. A very brief outline of this cytogenetic version of Lillie's hypothesis may be given as follows (ignoring all subsidiary aspects):

1. The only important forces of attraction operating during a mitotic cycle are those which obtain between any two homologous chromomeres. During the meiotic prophase it is this attraction that brings about the pairing of chromosomes.

Such an association in pairs satisfies the affinities of chromosome elements and when it has been attained, other elements, albeit homologous, are attracted no longer and are even repelled.

2. Every further mitotic activity is based on repulsion. The two charged centers repel each other and constitute the poles of the spindle. In so doing they orient the long molecules between them in parallel lines, such as are found in liquid crystals. Such lines or paths will transmit repulsions more effectively than unoriented protoplasm.

Pairs of kinetochores likewise repel each other and having charges similar to those of the centers, are repelled by the latter as well. Any given pair of kinetochores will fall into the lines of orientation between the centers because it is there that their repulsion activity meets the least resistance. The result of the reciprocal repulsions between the centers and the kinetochores is to force the latter into a position of balance or equilibrium—the metaphase plate. The spacing in the latter is not only due to the center-kinetochore repulsions but also a result of mutual repulsion of the bodies of the chromosomes (body repulsion).

The anaphase movement of daughter chromosomes is primarily due to the mutual repulsions of their kinetochores. A close approach to the poles becomes possible because there is a waning in the forces of repulsion of the centers.

It is obvious that the timing of the various processes as well as the forces of repulsion as they are represented in the centers, kinetochores, and chromosome bodies must be adjusted to each other or balanced in order to bring about an orderly mitosis. Hence Darlington has called this the "balance theory of mitosis." But brief examination will show that it makes no revolutionary changes in the older electrical hypotheses and the only novel feature lies in Darlington's claims regarding the numerical rules that control mutual attraction and repulsion of chromosome threads.

It is true, as Geitler (1938) says, that Darlington's assumptions concerning the details of mitosis are derived from the very phenomena that he is trying to explain. When chromosomes are able to approach the center it is because the forces of repulsion in the latter wane; when chromosomes become clumped despite body repulsion it is because there is a temporary weakening of such repulsion, and so on. Certainly this is no more than restating the observations in terms of the hypothesis, but that is exactly what almost every other hypothesis of mitosis has done. Darlington merely extends the method so as to give a more complete outline of the mechanism of the whole mitosis, and in so doing considers a more detailed series of steps. As a consequence the weaknesses of his hypothetical construction become more obvious than they would if he were dealing in the general terms of most other investigators of mitosis.

Aside from the descriptions of normal and abnormal behavior of chromosomes there is surprisingly little information that can be used as a sound basis for any electrical hypothesis. The primary

## HYPOTHESES OF MITOSIS

step of determining whether chromosomes and other cell elements are charged, and what the sign of that charge may be, has been taken by many investigators—starting with R. S. Lillie (1903) and culminating in the more recent experiments of Zeidler (1925), Botta (1932), von Lehotzky (1935), Kamiya (1937), and Churney and Klein (1937). Nearly all these workers have employed cataphoresis in attacking the problem, and there is little doubt that in such experiments the chromosomes behave as if negatively charged. Botta (1932) is convinced that the magnitude of the charge on the chromosomes may vary with the different phases of mitosis; if this can be confirmed it is manifestly of paramount importance in future considerations. It is of interest to note that Churney and Klein indicate that the nonchromosomal contents of the nucleus are electropositive, although it is possible that only the nuclear surface is thus charged. The cytoplasm, or at any rate its contained granules, according to Heilbrunn and Daugherty (1939) is also electropositive, though some work (von Lehotzky 1935) suggests that it is only very weakly charged or even isoelectric.

There would thus seem to be some physical basis for further considerations of electrical hypotheses. Without being hypercritical, one may point out, however, that cataphoresis involves very complex reactions and also that the passing of a current through a cell may bring about abnormal conditions. Bersa and Weber (1922), for instance, found an increase in viscosity of cells under the influence of an electric current, and it is needless to say that too strong a current may even cause death. Though the conclusions concerning the chromosomes, at least, are probably correct, it is desirable that more adequate controls be devised for experiments involving cataphoresis such as have been described here.

The circumstantial nature of much of the support for magnetic and electrical interpretations of mitosis is obvious, and most physiologists have taken a skeptical attitude concerning it. Thus Heilbrunn (1943) is inclined to be dubious about all magnetic hypotheses, because he believes that no one has as yet been able to demonstrate any effect of a magnetic field on mitosis. But apparently almost no work has been done on the effects of powerful magnetic forces on the chromosomes, possibly because of technical difficulties. However, such work as that of Ssawostin (1930) shows that the rate of protoplasmic streaming in algal cells is markedly affected in a magnetic field. This would at least bespeak an influence of magnetic forces on the cytoplasm and shows that further work is called for with respect to the nucleus.

Pease (1941, 1946) has argued against the importance of electric forces in the anaphase movement of chromosomes because under increasing hydrostatic pressure the chromosomes of the annelid Urechis are slowed and finally stopped (between 4,500 and 6,000 pounds). He reasons that since pressure tends to liquefy the medium surrounding the chromosomes, the latter should move faster rather than more slowly under such conditions, the assumption being that electric forces are not altered. But, although it is true that electric forces as such would be affected little by pressure, it seems very likely indeed that such high pressures may bring about at least temporary changes in various cellular elements which in turn might well influence their electrical conditions. The very common occurrence of multiple asters in cells recovering from lower ranges of such pressure would suggest that effects of this sort are possible, and it seems desirable that they be further investigated.

There is no question that many of the chromosomal maneuvers very strikingly argue for an electrical interpretation. No other forces furnish such a good explanation of the minor as well as the more obvious events of mitosis. But, by the same token, no other hypothetical consideration makes such a glaring exposure of our ignorance of the physical chemistry of the cell, for almost every step in such reasoning involves one or more assumptions.

### DIFFUSION

Electrical phenomena of a nature different from those just discussed are comprised in Teorell's studies of diffusion potentials (1937). These are set up, for instance, when an electrolyte is steadily diffusing across a boundary like a membrane, being continually replenished. But it would also operate if there is a steady diffusion from a certain part of the cell to another. Any electrically charged particle would be moved in such a system, positively charged particles being moved to the negative part of the diffusion potential field and vice versa.

As Teorell points out, we are here not concerned with any process like cataphoresis, for no current is being passed through the system. Instead, we are dealing with a type of propulsive force that is not ordinarily considered in chromosomal mechanics. It is conceivable that it plays a role in mitosis, for if a chromosome carries a charge it would be subjected to the influence of the system just like the particles. It is another question whether it would be moved at a rate comparable to that seen in some anaphases and it is not clear how the diffusion potential field would be set up within a spindle. In short there is here a suggestion for which there is at present no obvious method of application to the mitotic apparatus.

Rashevsky (1938, 1940, 1941), in attempting to lay the foundations of a mathematical analysis of cellular phenomena, attributes a major role to diffusion forces. The importance of such forces must indeed be very great and Rashevsky's mathematical considerations show how wide their range of influence may be. His endeavor to demonstrate their significance more specifically in the chromosomal movements of metaphase and anaphase is not altogether successful, however.

Rashevsky points out that if the mathematical conclusions derived from the division of the cytoplasm be applied to the division of the nucleus, one must keep in mind that the latter involves a more complicated system of structural elements. Perhaps the most important of these are the chromosomes and the chromosomal fibers.

Rashevsky's original hypothesis (1938) proposed that the chromosomes take their position in an equatorial plane because of repulsion-diffusion forces that originate from the two poles. The latter therefore represent diffusion centers. The chromosomes then become connected with these poles by elastic fibers. When the elastic forces exceed the diffusion forces just mentioned, as well as the forces that hold daughter chromosomes together, these chromosomes are separated and pulled to opposite poles.

In his considerations of 1940, Rashevsky does not deal with the establishment of the metaphase plate but makes a more detailed analysis of the anaphase movement. He argues that just as in the case of the cell as a whole, the diffusion forces in the spindle are manifested chiefly by its elongation (Rashevsky calls it the elongation of the nucleus, but his concern is really with the spindle). At metaphase the halves of a chromosome offer resistance to separation, and, since the chromosomes are connected through their elastic spindle fibers with the two poles, they thus serve to hinder and slow the elongation of the spindle as a whole. However as soon as the chromosome halves have separated from each other, this hindrance is removed, the elastic fibers contract and pull the chromosomes toward the poles, and the spindle is free to respond to the diffusion forces that tend to elongate it. Hence a curve of the rate of spindle elongation should show a sudden rise as soon as this point is reached.

There can be no question that the spindle of many forms undergoes elongation during anaphase, though this is not true of all species. But more specific data to test Rashevsky's theoretical conclusions are not available except for certain measurements made by Belar (1929a). These were made on the rate of elongation in spindles of the first spermatocyte division of the grasshopper Chorthippus (Rashevsky says "snail," but this must be a *lapsus calami*). Rashevsky concludes that Belar's data fit his theoretical curve very well indeed in that they show the sudden jump which should attend the separation of the chromosome halves.

There are several flaws in this argument, however. In the first place Belar procured measurements of changes in spindle length in only two cells (his data on second spermatocytes are concerned with the "Stemmkörper"). Further, he himself considers these measurements as very inexact since he could not be certain of the precise position of the centers. Finally, the point at which the curve shows its sudden rise does not, in reality, correspond to the stage at which the chromosomes leave the equatorial plate. Examination of Belar's data will show that this occurs much earlier. Indeed even the later anaphase, when even the chromosome ends have finally separated, is reached before this crucial point on the curve. In short, the data fit the theoretical curve because Rashevsky appears to have erred in their interpretation.

In his most recent paper on the subject, Rashevsky (1941) returns to the metaphase. Utilizing data given by Barber (1939) on the anaphase movement in second spermatocytes of Stenobothrus, he feels obliged to abandon the possibility that repelling forces are accountable for keeping the chromosomes in the equatorial plane. He concludes instead that the chromosomes maintain the metaphase position because they are subjected to the pull of the elastic chromosomal fibers exerted in opposite directions toward two poles. It is not until the daughter chromosomes separate from each other that they can respond to this pull, and this response constitutes the basis of the anaphase movement. In other words, the elastic chromosomal fibers are responsible for both the metaphase and the anaphase movement. However the data used in Rashevsky's calculations are again those of Belar (1929a), for Barber derived them directly from measurements made on Belar's photographs.[1]

It is clear that further and more exact measurements are needed in order to test Rashevsky's conclusions. Indeed, an exact timing of the observable steps may prove useful in more than one way (p. 79).

### STREAMING: CURRENTS

Spindle mechanics that involve regional currents and streams in the substance of the spindle have been suggested by many workers. Hypotheses of this sort are all the more natural since currents are demonstrably active in the division of the cytoplasm of the cell. But the suggestion already made by Berthold in 1886 that such cytoplasmic currents are directly responsible for chromosomal movement is almost certainly not tenable. The body of the spindle represents a special region in the cell that is structurally and physiologically differentiated from the cytoplasm, and currents in the latter do not flow directly into the spindle. Belar's (1929a) observations are in full support of such a conclusion.

Tischler (1922) was quite aware of the difficulties in applying to the spindle the conclusions reached for the cytoplasm. He made plain the very unsatisfactory state of our knowledge concerning

[1] Barber apparently did not realize that Belar (1929a) himself furnishes these data.

streaming phenomena as agents in chromosomal movement, but nevertheless expressed the conviction that streams, possible due to diffusion, must be present and must somehow contribute to the basic mechanism. That just about expresses the attitude of many of the more recent workers, although, as will appear, there is even now almost no factual evidence available. It is perhaps possible that there is a flow in astral rays and even in spindle fibers, but how this is related to the movement of chromosomes is not apparent unless one thinks of the spindle structure as differing from the ordinary conception of it.

The most complete development of an hypothesis involving streaming or currents in the spindle has been worked out by Schaede (1929, 1930) on the basis of his work on plants. According to this, the prometaphase involves the penetration of rather viscous, thin streams from the two polar caps into the spindle region. The physical basis for such streaming is not discussed, but there is no reason to doubt that the polar regions represent the seat of much activity. These streams propel the chromosomes before them into the middle region where a so-called indifferent zone is created at the meeting point of the two opposing sets of streams. It is here that the chromosomes come to rest and establish the metaphase plate.

A second factor is necessary to make the anaphase movement possible. According to Schaede this takes the form of an increase in surface tension of the middle region. That results in a lowering of the viscosity of the spindle substance in that region, which in turn brings about a streaming of spindle substance toward the poles where the viscosity is relatively higher.

Such a system of currents is correlated with the spindle structure, as Schaede conceives of it in the various plants he has studied. The spindle fibers as usually seen represent the narrow or lamellar streams of viscous protoplasm flowing from the poles to the equator. It is between these lamellae that the more liquid spindle substance of the midregion returns to the poles where its viscosity increases once more.

From prometaphase to telophase there are thus two sets of opposing currents, one toward and the other away from the poles (Fig. 16a). The chromosomes themselves are divided by surface tension which has to act exactly in the midline of each chromosome so as to induce displacement of chromosome material in opposite directions. The resulting daughter chromosomes then are transported in the flow of the more liquid protoplasm that is moving between the lamellae toward the poles.

The hypothesis solves one difficulty which has been a stumbling block to most other explanations (except that of Zirkle, p.

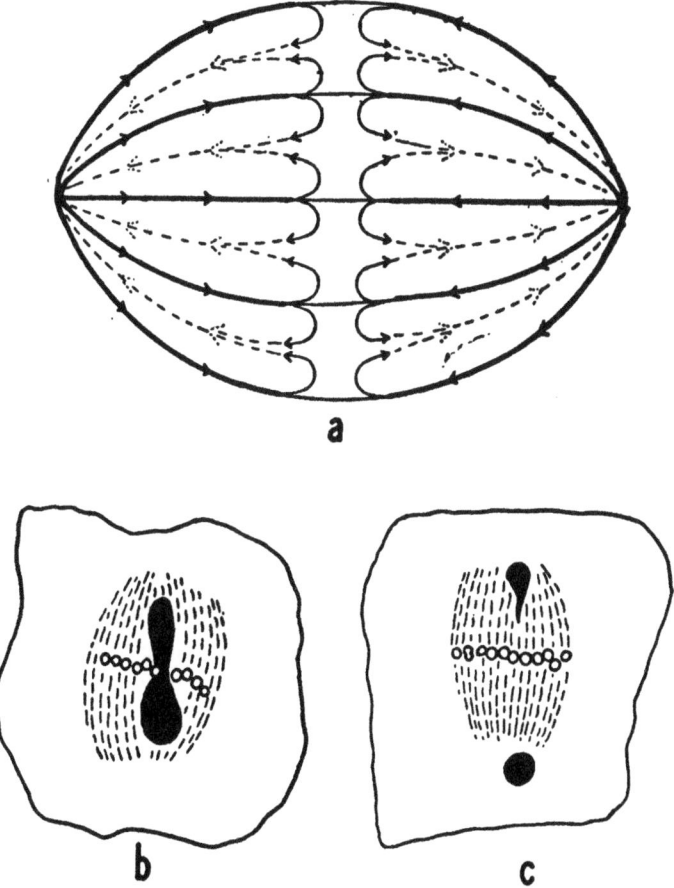

*Fig. 16.* a. Currents in the spindle, as suggested by Schaede (1931). Narrow streams of high viscosity (heavy lines) flow from the poles to the equator, and there is a returning flow of less viscous protoplasm (light, dotted lines). b. and c. Constriction and division of the nucleolus in the plant Cucurbita, while the chromosomes are still in metaphase position (Frew and Bowen 1929).

86). It accounts for the movement of such passive bodies as nucleoli and akinetic chromosomes. Especially among plants, many cases are known in which the nucleolus persists through metaphase and anaphase (Yamaha and Sinoto 1925, Schaede 1928, 1929, Zirkle 1928, Frew and Bowen 1929, as well as some earlier authors). In such species the nucleolus frequently migrates to the pole before the chromosomes have begun their anaphase movement. If at metaphase it lies exactly in the equator, it may even be divided, with the resulting two portions traveling in opposite directions (Fig. 16b and c). Manifestly there is here no possibility that the nucleolus is dragged along by the moving chromosomes, as has sometimes been claimed, for the latter have not yet begun their migration. A system of currents as postulated by Schaede would solve the difficulty.

Purely cytological considerations present some difficulties to Schaede's hypothesis. A lamellar or tubular form of spindle fiber is hardly indicated in most forms though it must be admitted that Schaede's figures of certain plant mitoses lend some support to his claim. It also must not be forgotten that Belar (1929a) too considered the existence of such lamellae as possible. However the poleward propulsion between such lamellae is conceivable only in a few species (Schrader 1931) in which the chromosomes are very short. In the case of long chromosomes with localized kinetochores the weight of evidence is against such a mechanism, for there the spindle fiber (that is, chromosomal fiber) can definitely be traced to the kinetochore instead of constituting lamellar walls on each side of the chromosome. Moreover the kinetochore always leads on the way toward the pole (with rare exceptions). All this would be difficult to explain if the movement were solely due to poleward currents flowing in long chambers.

Again, Belar (1929b) has objected that granules in the spindle of Tradescantia which evince Brownian movement nevertheless give no evidence of the existence of any currents. Schaede (1931) suggests that Belar was dealing with granules outside of the spindle, but both Ellenhorn (1933) and Becker (1935) have made observations which confirm Belar. It must be realized, however, that the gradual movement of such granules toward the pole would be difficult to prove or disprove, since it is almost impossible to keep a single granule within view for very long.

## HYPOTHESES OF MITOSIS

Difficulties would also be encountered in the physical basis postulated by Schaede for setting up the currents. To be sure, explanations involving differential surface tension correlated with cytoplasmic currents have been advanced with some success for the division of the cell as a whole (Bütschli 1876, McClendon 1913, Spek 1918). But it must not be forgotten that such cytoplasmic currents are much less complex than those suggested by Schaede for the spindle. Further, the animal cell usually is constricted sharply in the middle during division and no comparable deformation is to be seen in many plant spindles. Finally, the continuous cycle of alteration from a state of low viscosity at the equator to a more viscous state at the poles would present very special problems. Thus the liquefaction of substances demands pressure of a magnitude that is not easily attributable to surface tension and this as well as the reconversion to a more viscous consistency had better be postulated on a basis other than pressure changes.

To recapitulate, the mechanics for setting up a system of currents such as Schaede suggests are to a large extent hypothetical. If, however, a set of poleward currents be accepted as existing, no matter what their origin, one might well concede that they would play a role in anaphase movement, and certainly they would remove the puzzle presented by the mitotic behavior of the nucleolus.

### HYDRODYNAMIC FORCES

An extensive study of hydrodynamic forces has been made by Bjerknes (1902, 1909). The viewpoint is that of a physicist, without any reference to biological phenomena. An attempt to apply some of the findings of Bjerknes in an interpretation of the mechanics of mitosis was first made by the chemist Lamb (1907). His interesting paper was followed by those of Landau (1910), Woker (1920), and Cannon (1923) dealing with the same subject. We may agree with Lamb at the outset that we are dealing with "an ad hoc constructed hypothesis and intrinsically therefore only of hypothetical value." It is difficult to say, however, why so little effort has been made to put the hypothesis to an actual test.

The findings of Bjerknes that are pertinent to the questions here at issue may be outlined as follows:

1. Two spheres, pulsating synchronously in a liquid, that is,

expanding and contracting in a regular rhythm, attract each other when pulsating in the same phase and repel each other when pulsating in opposite phase.

2. Such spheres set up a system of hydrodynamic lines of force which resembles the disposition of the lines of force in an electric or magnetic field, and also an amphiastral configuration.

3. A neutral sphere suspended in the liquid is repelled by the pulsating spheres if it is lighter than the surrounding medium, and attracted if it is heavier.

If instead of pulsating, the two spheres oscillate, very similar fields of force are established. However the reactions are in a sense the reverse of those observed in the case of pulsating spheres, and correspond rather closely to those of an electric or magnetic field. In other words, if the two spheres oscillate in the same phase, they repel, if they oscillate in the opposite phase they attract, each other.

It will be seen that if no complicating features are considered, these facts furnish the mechanism that we require. The two spheres that pulsate or oscillate of course represent the centers or poles, and the suspended neutral sphere is the chromosome. Changes in the specific gravity of the latter, which quite conceivably occur regularly in the mitotic cycle, determine whether it is repelled or attracted. It is repelled to establish the metaphase, and in the anaphase movement is being attracted. But this simple outline may easily be complicated by other factors. Thus the pulsations of the polar spheres of the Bjerknes experiments may induce corresponding pulsations in the neutral sphere between them. How this would affect bodies like the chromosomes is not clear. Again, it is not inconceivable that the kinetochore, as a homologue of the center, is the responding element instead of the chromosome as a whole. This might serve in explaining the exact placement in the metaphase plate, for it must be remembered that the spindle spherule in the kinetochore has undergone division some time before the metaphase and that each of the chromosome halves carries a spindle spherule at prometaphase. If the chromosome has attained a position in the approximate middle, the spherule closer to one center might be induced to pulsate in the same phase as that center; its

## HYPOTHESES OF MITOSIS

sister spherule in the opposite phase corresponding to the other center. This would orient the chromosome exactly in the equatorial plane and when the chromatids finally separate, the pulsations of the spherules would bring about their movement to opposite poles without necessarily involving changes in specific gravity. The latter may of course occur also and thus aid the movement.

Other factors might complicate the picture still further but also furnish a basis for the great range of mitotic behavior in different forms. Thus the volume of the centers as well as their rate and amplitude of pulsation—or oscillation—would have a direct relation to the magnitude of the forces exerted. The response of the chromosomes would in turn be affected by the viscosity of the media in which they move.

At best, several difficulties remain. One is the configuration of the metaphase plate and the spacing of the chromosomes within it (Cannon 1923). The physical basis for this may be contained in the Bjerknes data though it is not obvious to the biologist. We must also reckon with the fact that each of the centers divides to provide the two poles for the succeeding division. If the system is such as has been discussed here, with the centers pulsating in opposite phase, each old center must give rise to two daughter centers which also pulsate in opposite phase. Again, how this can be brought about is not obvious to the biologist. The difficulty would of course be avoided if we are dealing with oscillatory poles, for there both are in the same phase throughout and no daughter center would have to reverse the phase of the old center. Finally it must be pointed out, as Cannon has already done, that Bjerknes states that a lighter body will be repelled from an oscillating body with a force that varies inversely as the 7th power of the distance between them. Thus an oscillating center would be effective over a very small range only, and it is a question whether it could bring about movement over the distances encompassed in the cell. This disadvantage is lessened in the case of a pulsating body since its effective force varies inversely with the square of the distance. But even this decrease in the effective force presents a serious difficulty for the hypothesis.

This brings us finally to the factual evidence. It must be conceded that this is of the slightest, resting as it does on the observations reported in only two papers. The first comprises the findings of Salazar (1925) who reported that in several types of fixed and stained cells in the rabbit the cloud of fine granules surrounding the center seem to undergo rapid alterations in distribution. The distribution changes from that of a hollow sphere to an oblong shape; Salazar ventures no more than to suggest that this indicates rapid expansion and contraction of the center, interior to the cloud of granules. This seems to occur even in cells that are inactive or, at any rate, not dividing. However, Salazar's evidence by no means proves that such alterations are rapid, and it is certainly inadequate to establish pulsation or oscillation in the ordinary sense of those words. The second piece of evidence indicates oscillation and was adduced by Huettner and Rabinowitz (1933) in the living cells of Drosophila. Referring to the centrioles they say: "They possessed a slight vibratory motion so that we could get both of them in focus at the same time for only a very short interval." Slight as it is, this observation may be of considerable importance. But the need of more extensive data is patent.

It must of course be realized that the existence of pulsation, even more than oscillation, may be very difficult to demonstrate. Centrioles are almost always close to the lowest range of vision and slight changes in their diameter might well be imperceptible to the eye. Again it is altogether conceivable that it is not the centriole but rather the centrosome or even a more indefinite polar region that pulsates. In either case, great rapidity and the absence of a sharp, delimiting boundary might render it almost impossible to recognize changes in volume.

In short, a hypothesis of hydrodynamic forces has some potential value. But just as when Lamb first proposed it, some factual support is very badly needed.

### TACTOIDS

Since about 1940 a growing number of investigators has endeavored to explain the events of mitosis on the supposition that the mitotic spindle is a liquid crystal or tactoid. The suggestion

## HYPOTHESES OF MITOSIS

that such might be the underlying nature of the division apparatus was originally made by Freundlich (1927) and elaborated in later publications (Freundlich 1937; Freundlich, Enslin, and Söllner 1933, Pfeiffer 1939, 1940). Tactoids arise spontaneously in certain inorganic solutions such as that of vanadium pentoxide. They are characterized by a parallel arrangement of long micelles or particles equally spaced in longitudinal lines, and representing a dynamic equilibrium in a system of interionic forces. The typical cusping that at once suggests the mitotic spindle is due to surface tension forces. In 1941 Bernal and Fankuchen demonstrated that in tobacco virus there may be a formation of organic tactoids.

In this connection it is pertinent to mention the "protein spindles" or "protein crystals" that occur in the cytoplasm of several plants; these were observed by Hartig in 1856. The more recent studies of Küster (1935, 1948), Weber (1940), and others have shown that these structures resemble both tactoids and spindles in usually being fusiform and exhibiting a marked birefringence. They differ from tactoids in attaining a size that frequently compares favorably with that of spindles and in having a stability that permits bending and distortion without consequent disintegration. A renewed study of the physical chemistry of these "protein spindles" may well prove to be very rewarding in throwing light on the nature of the orientation forces of both tactoids and spindles.

A more detailed and specific effort to explain the anaphase movement of chromosomes on the basis of our rather vague knowledge of tactoids was made by Bernal in 1940. He pointed out that there is some evidence that if in a tactoid there is included some of the unoriented medium, the latter will also assume fusiform shape. Such enclosed formations ("negative tactoids") will place themselves parallel to the long axis of the original "positive tactoid" (Fig. 17).

Making certain assumptions without adhering too closely to our cytological information concerning the spindle, Bernal then outlines the course of the anaphase as follows:

1. The two centers would bring about the formation of a positive tactoid—the spindle.

2. Each pair of daughter kinetochores, still in association, would induce the formation of negative tactoids within the positive tactoid.

3. The daughter chromosomes move apart because the whole system of tactoids, with which they are associated, is subject to elongation—the anaphase movement.

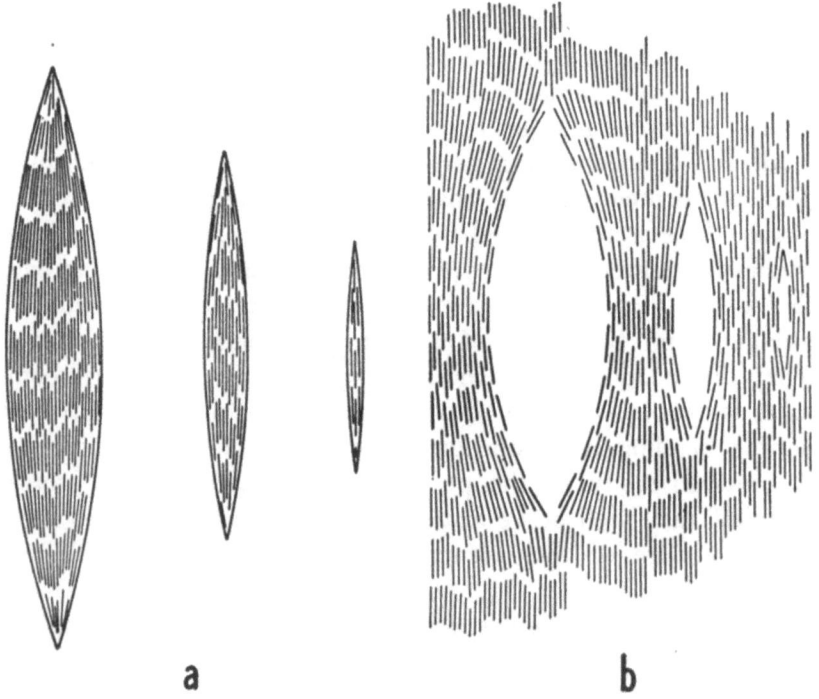

*Fig. 17.* a. Positive tactoids. b. Negative tactoids (Bernal 1940).

4. The long particles of the tactoid break up into shorter ones and the whole structure becomes disoriented—the telophase.

As Bernal says, the evidence for the truth of this picture is still very scanty, and every cytologist is bound to agree with him. For instance, the actual conditions of association between chromosomes and fibers are difficult to bring into such a conception. Again, though the spindles of some forms do elongate in anaphase, there is no question that some others do not. At best it is not clear how elongation alone could bring the chromosomes closer to the poles (Barber and Callan 1943).

## HYPOTHESES OF MITOSIS

Although this highly speculative explanation thus meets with decided difficulties, there are certain findings which forbid too abrupt a discarding of the tactoid hypothesis. These are best presented in the publications of Östergren (1945a and b, 1949a, 1950a and b, 1951), and are particularly interesting where they pertain to the establishment of the metaphase plate. This important phase constitutes one of the most obscure points in the analysis of mitosis and receives very little consideration in most hypotheses.

Östergren envisions the process as follows: the transport of the chromosomes into the mid-region between the two poles or centers constitutes the first step. In this he follows the hypothesis of Drüner (already discussed on p. 66) in which the equatorial position of the chromosomes is considered to be the resultant of opposing pulls from the two poles. Unlike Drüner, however, Östergren, in his more recent publications, believes the effective traction to be due to tactoid forces (p. 74). The next step is the more precise arrangement of the chromosomes in the metaphase plate, and this also involves tactoid forces. The tactoid comprises a system in which mobile particles are in a dynamic equilibrium. There would be a tendency to eject bodies such as chromosomes laterally out of the system. The resultant of this ejecting force and the forces of poleward traction determines the position of the chromosome in the metaphase plate (p. 67).

The anaphase movement of the divided chromosome finally would involve the tactoid forces already active in providing a poleward traction during prometaphase.

There are several objections to Östergren's explanation. Thus, nucleoli and chromatoid bodies if present are usually found within the body of the spindle (Fig. 16), whereas, having no connection with the poles, they should be shunted out of the spindle with great regularity. Again, large chromosomes are by no means always placed at the periphery of the plate nor does the spacing of the chromosomes with respect to each other receive an adequate explanation. It is possible that these objections can be overcome through subsidiary assumptions, and they may be ignored for the present while we consider criticisms of a more general type. Thus Hughes (1952) points out the following differences between tactoids and spindles:

1. Tactoids lack orientation centers or centrioles.
2. Tactoids are exceedingly labile while spindles are not.
3. The internal structure of tactoids is uniform whereas that of spindles is not.
4. Unlike spindles, tactoids can not elongate without increasing in girth.
5. Tactoids grow readily in size whereas spindles do not.

The last of these criticisms is not a valid one since the division apparatus of many species, for instance, among the Amphibia and Hemiptera, may start as a miniature spindle in the preceding telo-

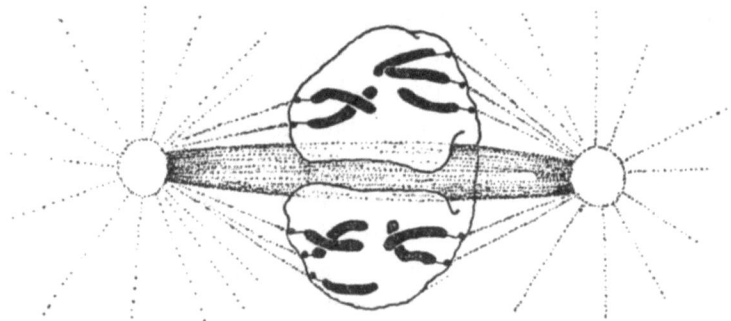

*Fig. 18.* Spindle conditions during division in protozoa such as Barbulanympha and Trichonympha. Semidiagrammatic drawing of conditions shown by Cleveland (1934, 1949b).

phase and increase in size until the full dimensions are reached at metaphase. However, the remaining criticisms make it clear that if the mitotic spindle is to be regarded as a tactoid, our concept of the tactoid must undergo various modifications.

But difficulties arise also from certain embryological and cytological findings. The cases at issue, which by no means constitute a complete list, are discussed at greater length by Schrader (1951) and may be given here very briefly:

1. In monopolar spindles there is only a single focal point and the fibers are spread widely at the opposite extremity (Metz 1933, Scott 1936). (Fig. 19.)
2. In some cases, single chromosomes may be separated from the main spindle, taking with them parts of the spindle, that is, their chromosomal fibers. In other instances the metaphase spindle

## HYPOTHESES OF MITOSIS

is forcibly split or spread at the equator through a lateral shift of a whole group of chromosomes (Schrader 1946a and b, Hughes-Schrader 1948b). (Fig. 13.)

3. In certain Protozoa the nucleus never breaks down completely during division and is interposed as a partial obstruction to any orientation that may obtain between the two poles (Cleveland 1934). (Fig. 18.)

4. Two spindles that intersect each other in their mid-regions are sometimes encountered in embryological experiments on echinoderm embryos (Baltzer 1908, 1911).

No ordinary tactoid could exist in the various forms shown by the spindles in the cases just given. It is, as already pointed out, a very labile structure which disintegrates readily when any agency acts so as to split or otherwise deform it. Some of these findings (Cases 1 and 2) might be explained on the basis that the mitotic spindle really represents an aggregation of smaller tactoids instead of a single, large one. That would certainly have to be the supposition in a case such as Acroschismus (Fig. 7). But this in itself would call for assumptions that are not encompassed in the ordinary understanding of tactoids—for instance, assumptions concerning the means for preventing a fusion of tactoids when they are massed together. At the very least, the various objections just discussed make it obvious that the tactoid hypothesis can not be applied in its unmodified form.

Indeed, the necessity for some modification must follow from any careful consideration of our cytological data. It should not be forgotten that if a tactoid hypothesis is to be applied to the mitotic mechanism, we must take into account two powerful factors that are not present in the primary liquid crystal. These are, of course, the centers and the kinetochores. If these can be built into a tactoid system, as recent workers assume, the tactoid might at once gain greatly in stability. It would then be enabled to broaden its range of behavior, and thereby furnish an answer to all the objections listed above. The difficulty here lies in the fact that our knowledge of tactoids is still rather limited and we may be making assumptions that are entirely unwarranted.

It may also be objected that in such a modification the tactoid hypothesis comes to take a rather subsidiary rank, one in which its only contribution to our analysis of mitosis lies in its emphasis on interionic forces. Hughes (1952) voices some such objection in a slightly different connection: ". . . if the tactoid hypothesis is to be whittled down to such vague generalities, it becomes doubtful whether it is worth having." However, as I have stated elsewhere, the final solution of the mitotic problem will almost certainly make use of parts of several different hypotheses, and none of these can be considered as having failed if it has contributed to the final answer.

### CHROMOSOME AUTONOMY

At intervals, especially during the last twenty years, the opinion has been expressed that the mitotic movements of chromosomes are largely autonomous. This implies that the forces involved reside or originate in the chromosomes themselves and that therefore the spindle apparatus plays a subsidiary role (see for instance Frey-Wyssling 1938). So far as the general problem of the mechanics of mitosis is concerned, such a view serves to emphasize that the role of the chromosomes is perhaps not as passive as is often assumed. However the analysis of such chromosomal forces is peculiarly difficult, and their nature is usually not even guessed at. The only concrete suggestions have been advanced by Metz (1933, 1936) and Swann (according to Hughes 1952). According to Metz we are concerned with progressive viscosity changes called out in the neighboring medium by the individual chromosome itself (p. 82). Swann believes that some substance extruded from the kinetochore affects the orientation of the particles in the chromosomal fiber and thereby initiates the movement of the chromosome to the pole.

Like Metz, the other proponents of a chromosomal autonomy in mitotic movement have nearly always been induced to adopt such a hypothesis through the study of special cases which rule out any explanation based on contraction, expansion, streaming, or electrical forces. In Sciara (Metz, Moses, and Hoppe 1926) the first spermatocyte division is monopolar. There is no formation of bivalents and the full complement of univalents lies rather close to the single center. In the ensuing mitosis, one set of homologues—

that derived from the father—leaves this aggregation and travels to the opposite side of the cell on lines radiating from the pole. The rest of the chromosomes approach this pole, from which they

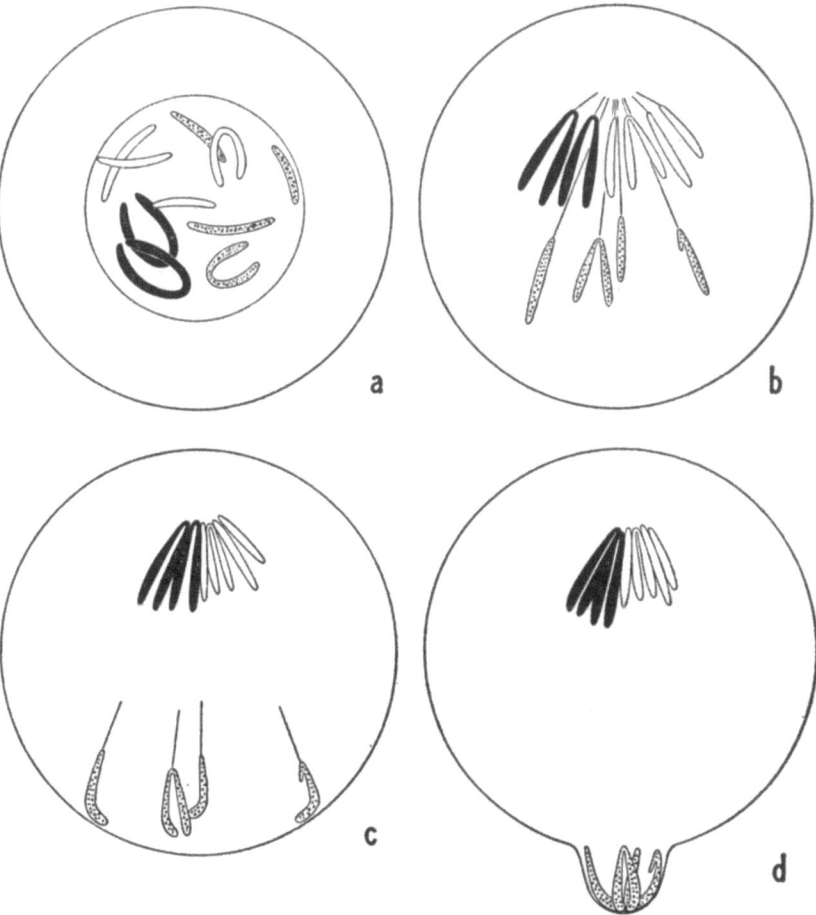

*Fig. 19.* Schematic presentation of the first spermatocyte division in the fly Sciara (Metz 1933). Paternal chromosomes are stippled, maternal chromosomes are white or black. a. Stage corresponding to metaphase. No equatorial plate is formed and there is no pairing of chromosomes. b. Early anaphase. Maternal chromosomes move toward the single pole; paternal chromosomes move away from it. c. Late anaphase. Maternal chromosomes have reached the pole; paternal chromosomes are reaching the opposite periphery of the cell. d. Telophase. Paternal chromosomes gathered in a bud which is cast off.

are separated by only a short distance. So far as the mechanism of this mitosis is concerned, by no means the least puzzling part lies in the fact that the angles of all the chromosomes are pointed at the single center throughout this anomalous movement. This indicates that the kinetochores are oriented toward this center even when the chromosomes move away from it. Indeed the spindle fibers that connect the chromosomes with the pole seem to function as a restraining influence, and to all appearances the movement to the distal part of the cell occurs despite, rather than because of, them.

The case presents difficulties for all mitotic hypotheses thus far devised. There is no formation of bivalents, and yet homologues separate from each other. Moreover, the segregation is not at random, for the chromosomes that move away from the pole (and finally degenerate) are always the paternal chromosomes. The kinetochore may be functioning, but the forces that preponderate and decide the movement lie outside of the kinetochore. It is easy to see why Metz was driven to seek the decisive factors within the individual chromosome itself, for any agency residing elsewhere in the cell would be difficult to visualize (Fig. 19).

Nor can this case of Sciara be set aside as an enigmatical exception, for contrary to general opinion certain of its features are paralleled in other species. In the first spermatocyte of the beetle Micromalthus (Scott 1936) all the chromosomes of the haploid male move away from the single pole, while their kinetochores continue to be oriented toward it. Like Metz, Scott concludes that this movement is autonomous. In the coccids Gossyparia, Pseudococcus (Schrader 1923, 1929), and Phenacoccus (Hughes-Schrader 1935) the meiotic mitosis is very similar to that of Sciara and in the two last-named there is no trace of a formation of bivalents. Although the size of the chromosomes and the conformation of the kinetochores in all these Coccidae make it difficult to determine their orientation with respect to the pole, the mitotic movement clearly is very closely akin to that of Sciara.

Evidence in support of chromosomal autonomy is also present in much of the experimental work on echinoderms. Thus the chromosome halves in conjunction with monasters will separate

## HYPOTHESES OF MITOSIS

readily, though the distance between them remains small (Wilson 1901, M. Boveri 1903, and others). However in the experimentally induced monasters of the annelid Urechis, Belar and Huth (1933) showed that the daughter chromosomes move considerably farther apart than is usual in corresponding figures of echinoderms. In fact in some of Belar's drawings this movement approaches that seen in normal anaphases. The evidence thus is adequate to show that the daughter chromosomes may separate for some distance quite independently of a regular bipolar spindle apparatus. But evidently some polar activity is necessary for any movement that goes beyond the primary step of separation, and if a subsidiary pole is formed this movement at once comes close to a normal one. The anaphase separation in certain plant chromosomes that have strayed from the regular spindle (Juel 1897, Belling 1927, Suita 1939) may also be more or less normal, but then it is always associated with an independent miniature spindle whose derivation is obscure.

All of the preceding comments pertain to an anaphase movement that, albeit indicating a considerable autonomy of the chromosome in the mitotic mechanism, is nevertheless associated with some part of the normal spindle apparatus. Entirely apart from this is the process through which the chromosome divides internally, or rather duplicates itself. Almost nothing is yet known of the mechanics involved, but certainly it occurs quite independently of any spindle activity and thus is truly autonomous. When this process is completed the daughter chromosomes may "fall apart" even when no spindle is present, and such seems to be the case in the experiments of F. R. Lillie (1906), who found that chromosomes of Chaetopterus would multiply even when the amphiaster was prevented from forming. It occurs normally also in endomitosis, where polyploid numbers of chromosomes are attained without the intercession of a spindle (Geitler 1937, Berger 1938, Bauer 1938, Painter and Reindorp 1939).

But in orthodox mitoses this falling apart of daughter chromosomes may be initiated by a kinetochore activity which also argues for a certain amount of autonomy. Thus in the case of certain mantids, Hughes-Schrader (1943b) finds that such kinetochore movement unquestionably occurs before the spindle is formed and without reference to the poles. Swanson's conviction (1942) that such kinetochore activity is possible only when the chromosomes are on the metaphase plate may be justified in many species, but certainly there appear to be cases where the kinetochores behave quite autonomously.

In some forms the kinetochores do not move in advance of the chromosome body in the initial anaphase step. Thus in the grasshopper Chorthippus (Belar 1929a) the daughter chromosomes of the second spermatocyte undergo about one third of their anaphase movement to the poles without any precession of the kinetochore, and the movement is even more extended in the corresponding division of the amphibian Amphiuma (Schrader 1936). This would therefore bespeak an autonomy on the part of the chromosome body as well, for though it occurs in the spindle it is not obvious how the latter could bring about such a movement. In many other species the case for an autonomy of the chromosome body is not so convincing, for the clear line which often separates daughter chromosomes as anaphase begins may represent nothing more than a new formation of pellicle between them. It may incidentally be remarked that the presence of an extensive pellicle might explain the separation between the chromosomes in the metaphase plate (Metz 1934), though of course it would not easily account for their disposition in particular patterns.

An hypothesis involving what he calls chromosomal autonomy has been sponsored also by Bleier (1930a and b and 1939), who, like Metz, was almost forced to take such a point of view by the independent behavior of chromosomes in certain plant hybrids. He believes, however, that the chromosome can function only in a medium of nuclear or spindle substance (Paragenoplastin) and that the latter has a mitotic cycle of its own which in an orthodox mitosis is coordinated with that of the chromosome (see also Koerperich 1930).

Bleier, like many other investigators (Vandel 1928) believes that an orderly and normal anaphase movement is contingent on the presence of chromosome elements in apposition, be they pairs of homologous chromosomes in meiosis or chromosome halves in any other mitosis. The primary anaphase mechanism then is one of mutual repulsion between the members of such a pair, whatever the physical nature of these repulsive forces may be. The erratic movement of univalents thus is due to the very fact that they have neither "divided" internally nor been able to pair with a homologue.

For the sake of argument one may ignore the cases like Sciara and the Coccidae where an effective anaphase movement occurs without such pairing. It may only be pointed out here that the autonomy that Bleier claims for the anaphase movement is no more than that which Lillie sponsored in his electrical hypothesis, even though Bleier does not commit himself concerning the type of force that is involved. Nor does he enter into the question of the establishment of the equatorial plate, and it is not obvious how mutual repulsion can bring about this most difficult of mitotic phases. Finally, it now is clear that the spindle fiber apparatus is a reality and that, unlike Bleier, one must take it into account in any mitotic hypothesis.

But that is not to doubt the very great influence that the chromosome may have in the mechanism of mitosis. Thus the great problem of segregation in the meiotic divisions is indubitably based to a very large extent on properties residing in the individual chromosome itself. The segregation of the unpaired X and Y in Melophagus (Cooper 1941a); the unified behavior of compound sex chromosomes (Payne 1909); the orderly alternate separation of chromosomes in the translocation chains of Oenothera (Cleland 1926); the preferential segregation in trisomic Drosophila (Sturtevant 1936); the separation of the paternal from the maternal chromosomes in Sciara (Metz 1936), and many similar cases all pose problems that, albeit still obscure, nevertheless bespeak a most important and active role for the chromosome in the processes that determine segregation.

The sum and substance of all these considerations amounts to the following: The chromosomes themselves play an active part in the mitotic mechanism. Any movement beyond the initial separation is, however, associated with the presence of some part of the mitotic apparatus, even though the relationship is not always obvious. The forces that are involved in such an active participation of the chromosome are still obscure, although the proponents of chromosomal autonomy have performed no small service in insisting that the chromosome is not a mere passive body which is transported hither and yon by forces more or less extraneous to it.

# IV. Related Problems

IT HAS BEEN SAID that the complete analysis of mitosis in a single species would give answers to all our questions and that the apparent skipping about from one species to another, in which we have nearly all indulged, has often befogged the issue. There is some justice in this statement. A successful exposure of the mechanism that underlies one complete mitotic cycle would no doubt solve most of the others also, for it may well be granted that a basic principle is common to them all.

But there are practical difficulties in attaining the goal in this way. It may be stated without danger of contradiction that in no one species are the main steps of mitosis all shown with unequivocal clarity. Technical difficulties or peculiar special conditions always tend to obscure certain of them. But in another species these very stages may present no such hindrances to study, and when this is known it becomes very hard not to appeal to them. Indeed as will appear in the following pages, it may actually be better tactics at the present time to center an attack on a given stage of mitosis and seek it out in whatever forms it may be most open to analysis, instead of aiming at a solution of the complete mitotic cycle at once. The arbitrarily delimited steps of mitosis are themselves complex and the most helpful, albeit undeniably laborious, method would seem to lie in dissociating each step into its component processes.

As an example of an attempt thus to dissect a single step may be mentioned Ris's work on the anaphase movement of the chromosomes of the aphid Tamalia (1943). By carefully timing the progress of the chromosomes to the poles in living, normal cells, he was able to establish that in Tamalia this seemingly obvious and most straightforward of mitotic processes always involves two movements separated by a brief delay. This would signify that at least two factors are comprised in it, as Ris points out (p. 79).

Belar's attempt (1929a) to make such an analysis on the basis of the timing and measuring of photographs failed to some extent because he could not be sure of the location of the spindle poles and also because his various stages are too far apart.

If the complexity of the mitotic mechanism is once conceded, an examination of its minutiae immediately becomes significant. In the following pages a survey of problems that are usually considered only incidentally as entering into mitotic questions will be given. Without doubt they are part and parcel of the mechanism of the cycle even though it is not clear at the present time how importantly they bear on it. Very often it is during meiosis that certain features stand out most clearly and hence many of our best data stem from that period of germ cell development. In making such a survey the terms "attraction" and "repulsion" are extremely useful and they are employed here to denote the approach and separation of cell elements without any connotation of the forces that may underlie such movements.

### RESTING STAGE

For the cytologist, the most obscure period of the mitotic cycle is the resting or interphase. Everybody now seems agreed that it is anything but quiescent and the old term stems, of course, from the cytologists of former years who could see in it nothing of the dramatic maneuvers that characterize later stages. The cytological difficulty lies in the fact that the outlines of the chromosomes are optically diffuse and vague at this time and that therefore their movements cannot be followed with certainty, if at all. Indeed even now, when we are convinced that much of importance transpires in the cell during this period, we know almost nothing in it that pertains to mitosis (see the review by Becker 1938). Some recent cytochemical analyses indicate that it is during the interphase that the normal amount of chromatin (as measured by the DNA) is restored after being halved in the preceding anaphase (p. 53). If this proves to be correct, then the interphase is a period of synthesis for at least one of the constituents essential for mitosis.

# RELATED PROBLEMS

## PAIRING

The leptotene, zygotene, and pachytene stages of meiosis present several maneuvers that are probably of great importance in the later parts of the cycle. The phenomenon of meiotic pairing, which involves a very precise apposition of homologous chromosomes—chromomere to chromomere, or perhaps even gene to gene—represents one of the most puzzling phases in the maneuvers of the chromosomes. The modern tendency to regard it as involving no special forces that are not present also in an ordinary mitosis, but rather an alteration in timing and emphasis, is unquestionably more promising of a solution of the problem than were the older views. But no practical hypothesis concerning the actual forces involved has yet been proposed. It will be realized that Darlington's rule of attraction in twos simply expresses a supposed fact, by no means accepted by all cytologists, and does not represent an explanation of the underlying mechanism. Actually, even this much of a generalization has now been abandoned by virtually every cytologist: the chromosomes are not undivided, single strands when they come together in synapsis (Huskins 1937, 1942; Geitler 1938; Nebel 1939).

We are thus further than ever from solving a problem which in the last analysis is one of physical chemistry. Unfortunately the physical and chemical interpretations are at a very tentative level and have made as yet no contact with the cytological and genetical findings. The skillfully constructed hypotheses of Friedrich-Freksa (1940), Delbrück (1941), Fabergé (1942), and Jehle (1950) have been advanced with all due awareness of this gap that remains to be bridged and of course constitute no more than first attempts toward a solution.

The old question of the chiasma—whether or not it signifies that an interchange of sections has occurred between pairing chromosomes—is being debated once more. That an interchange of chromosomal regions may take place has been sufficiently established (Stern 1931, Creighton and McClintock 1931), but the mechanism that underlies it is still not known. The universality of the interpretation given to the chiasma by Darlington has been seriously questioned by Matsuura (1937, 1949b), Hughes-Schrader (1943a and b), and Cooper (1945, 1949). In working on the problem much difficulty arises from the fact that it is often almost impossible to be certain that a true chiasma is present in a given case. Certainly, either alternatively or in addition to the chiasma, other agencies may also be active in holding a meiotic tetrad together during the late prophase and metaphase, for instance, "collochores" (Cooper 1944), terminal attraction (see under "Telomere"), unanalyzed pairing forces (Hughes-Schrader 1943a and b), and heteropycnotic attraction (Schrader 1941c).

The pairing behavior of the sex chromosomes in certain insects may involve special problems. For instance in certain Hemiptera the pairing of the X and Y chromosomes is restricted to a momentary contact just before the second spermatocyte division—the "touch and go" pairing of Wilson (1925). Perhaps this is due to nothing more than spindle forces that move both chromosomes to the same locality (Darlington 1939b) but in Rhytidolomia the contact between X and Y appears to involve a certain region of each chromosome and thus bespeaks an actual attraction (Schrader 1940b). Perhaps just as puzzling are the cases in certain plants (Lorbeer 1934) and some Neuroptera (Naville and De Beaumont 1933), where the chromosomes do not even approach each other but nevertheless take a vis-à-vis position in the spindle. This has been called "distance conjugation" by Lorbeer.

### TELOMERE

The assignment of special functions to the ends of the chromosomes—so strongly indicated by the bouquet stage—has been sponsored for some time on genetic grounds by Muller (1941). He has proposed the term "telomere" for the terminal chromomeres which seem to be especially concerned. Indeed, important and localized properties in the chromosome ends are also attested by such special cases of terminal adhesion and attraction as reported by Schrader (1940a), Hinton and Atwood (1941), and Hinton (1945). Similarly, Darlington's conceptions (1937) of "terminal chiasmata" and "terminal affinity" evidently recognize special activities in the end regions.

The custom, so generally adopted by cytogeneticists, of interpreting all terminal unions in bivalents as signifying chiasmata is not justified. A coming together of chromosome ends that is demonstrably independent of chiasmata has been observed by Taylor (1949) in the live cells of Tradescantia, and the terminal connections in the bivalents of certain forms of Trillium are considered by Matsuura and Haga (1942) to be similarly independent of any chiasmatal union.

A special behavior of chromosome ends that is not dependent on any permanent, structural element has been reported by McClintock (1938, 1939). She found that when maize chromosomes are broken during meiosis there is a strong tendency for the broken ends to come together again and fuse. But in the embryo the broken chromosomes do not undergo such fusion. It is obvious that the region where a break has occurred is endowed with special properties, and these may be affected by the surrounding medium. It is possible that the behavior of broken ends does not involve the same problem that is presented by uninjured chromosome ends. However, in both cases it is only a sharply delimited terminal region of the chromosome that is concerned and never a lateral portion.

Altogether, the evidence is more than sufficient that chromosome ends are endowed with special properties which may well exert a considerable influence on the behavior of chromosomes.

Attempts to identify the telomere with some structural feature in the chromosome have not been too successful. However Warters and Griffen (1950) report that the salivary chromosomes of certain species of Drosophila terminate in visibly specialized chromomeres which they believe to represent the telomeres. They describe them as globules with a dark shell and clear interior.

#### HETEROPYCNOTIC ATTRACTION

Certain features of the prophase in some species reflect rather special conditions of chromosome structure. Thus the attraction between parts or whole chromosomes that are heteropycnotic is probably not due to forces identical with those that bring about the exact chromomere-to-chromomere apposition of synapsis. For one thing, it appears to be nonspecific in that heteropycnotic chromatin tends to aggregate during a certain period (see review by Schrader 1941c). Usually this attraction is of short duration and in most cases has disappeared before metaphase. Some striking illustrations of this behavior have been reported by Slack (1938) and Schrader (1941d) but nothing can be said concerning the forces involved in such attraction. However it is becoming generally recognized that there may be several types of heterochromatin which, albeit cytologically indistinguishable, differ fundamentally in their physiological processes.

### KINECHORE ATTRACTION

In certain cases there is evidence that there may be a similarly nonspecific attraction between the kinetochores during meiotic prophase (Matsuura 1949a, Matsuura and Kurabayashi 1951). Darlington (1937) suggests that this also may be the case in the plant Agapanthus and the insect Stauroderus. Perhaps such attraction is also active in cells with polytene chromosomes in certain Diptera (Frolowa 1936), but no direct, decisive observations are available on this point. It is possible that in some cases the attraction between kinetochores or between kinetochores and chromosome ends may actually involve heteropycnotic attraction, for in many instances the region of the chromosome adjacent to the kinetochore is heteropycnotic.

The detached kinetochores described by Pollister (1939, 1943), in molluscs, aggregate around the centriole, but there is of course no means of knowing how the special conditions evidently obtaining in oligopyrene spermatocytes affect the various elements of the nucleus. At any rate an attraction between kinetochores that is hinted at in these cases would of course indicate a complete reversal in the action of forces controlling them during prophase and later during metaphase and anaphase.

### THE NUCLEAR MEMBRANE

It will be evident from what has been said in the present treatise that the nuclear membrane plays a very important role in the mitotic cycle. Thus during the long period from telophase to metaphase it is interposed between centers and chromosomes, and their whole range of behavior must continually reflect their interaction not only with each other but also with the nuclear membrane. Unfortunately very little is known of its structure and physical chemistry. Doubts of its existence, so frequently expressed in former years, can however be entertained no longer. In almost any organism the phase-contrast microscope now gives the final proof which was already implied in investigations such as that of Chambers and Fell (1931).

The nuclear membrane is evidently very plastic, as is attested by the distortions which it may undergo in the ordinary prophase maneuvers (Fig. 12), as well as under the pressure incident to centrifuging and other experimental treatments (Luyet and Ernst 1934). The electron-microscope studies of Callan, Randall, and Tomlin (1949) led them to conclude that the nuclear membrane of amphibian oöcytes consists of two distinct layers, of which the outer one shows a porous structure. Its permeability varies in different organisms and even in different tissues for, whereas the membrane of the amphibian oöcyte just mentioned permits the passage of salts and sugars (Callan 1949, 1952; Abelson and Duryee 1948), the epithelial nuclei of Amphibia as well as the germinal vesicle of echinoderm eggs (Beck and Shapiro 1936, Churney 1941) are permeable only to water and not solutes. The well-known difficulty of vitally staining nuclear structures is no doubt due to the interposition of the nuclear membrane in uninjured cells, though Monné (1935) reported that with certain dyes the nuclear contents are stained rather faintly. However, such a stain is only transitory and the color quickly disappears again. More information concerning the chemistry of the nuclear membrane is badly needed and findings such as those of Churney (1941) can be regarded as no more than a promising beginning.

The ease with which certain nuclear constituents pass through the nuclear membrane out into the cytoplasm is surprising. There is much cytological evidence on this score. A number of earlier investigators concluded that fragments of the nucleolus or droplets of nucleolar material regularly make their way through the membrane to participate in the chemical activities that go on in the cytoplasm. This was frequently doubted though it now appears that many of such observations were correct. In the ovary of certain Crustacea and Hemiptera (Lison and Fautrez-Firlefyn 1950, Schrader and Leuchtenberger 1952) large droplets of nucleoproteins are extruded through the nuclear membrane, and in the Hemiptera the entire nucleolus may pass through as well. Similar evidence for a ready penetrability is furnished in Sciara, where Berry (1941) showed that an entire chromosome in a rather condensed condition may pass through the membrane. The fact that such large masses of material may be extruded—apparently in the normal course of events—without injury or visible alteration of the nuclear membrane bespeaks very special properties for it.

### THE PRE-METAPHASE STRETCH

A peculiar but significant phenomenon was noted by White (1941) in the meiotic prophase of certain mantids, and carefully analyzed by Hughes-Schrader (1943b). The peculiarity lies primarily in a sudden movement of the two kinetochores of the bivalent towards opposite poles, contemporaneously with the breakdown of the nuclear membrane and the formation of the spindle. In the process the bivalents are stretched open with an intensity which varies widely among different species but is often so extreme that the chromosomes become markedly attenuate and the separation of the two kinetochores may extend over two thirds of the length of the spindle. An elongation of the spindle accompanies this movement in most species, but it is responsible for only a part of the separation of the kinetochores.

All this occurs prior to metaphase. In attaining the latter stage the stretched regions of the chromosomes contract (coil?) again, once more bringing their kinetochores closer together; meanwhile the bivalents, maintaining their bipolar orientation, move to the equator. The processes of stretch, recontraction, and congression at the equator may overlap, for the chromosomes react asynchronously, and one bivalent may thus be moving toward the equator while the kinetochores of another are still moving toward the poles.

The stretch itself might be attributed to a strong mutual repulsion between homologous kinetochores, especially since in some species such a repulsion is indeed evinced before the nuclear membrane breaks down and before any spindle fibers are visible. This initial separation of kinetochores occurs without orientational reference to the centers. The bipolar orientation during the ensuing stretch stage might then be attributed to the longitudinal structure of the spindle body permitting movement only parallel with the interpolar axis. That such is not the case—and that the stretch phenomenon involves primarily an attraction between kinetochore and center—is attested by the facts that the kinetochore of a univalent (e.g., the unpaired X of certain species) orients toward the nearer pole, and that chromosomal fibers are frequently demonstrable in both bivalents and univalents. Finally, the behavior of the sex trivalent ($X^1X^2Y$) of many species bespeaks a kinetochore-center attraction. It has opened out prior to the stretch into a chain of three chromosomes whose kinetochores are so widely separated that a specific influence on one another is difficult to envision. Moreover, as the spindle forms and the stretching of the bivalents is initiated, the kinetochores of the sex chromosomes react individually each to the pole nearest it, instead of invariably coorienting so that the two X's are opposed to the Y as would be expected from a repulsion between homologous kinetochores. The resultant, not infrequent malorientation of the sex trivalent during the stretch is almost always corrected before anaphase, showing that a kinetochore, even after the formation of chromosomal fibers, may change its orientation from one center to the other, and that in the process a whole chromosome may be shifted across the equator (Hughes-Schrader 1943b, Inamdar 1949).

The course of meiosis in these mantids thus presents many points significant for mitosis. The primary separation of kinetochores may occur independently of the rest of the spindle apparatus. An attraction between kinetochores and centers appears to be clearly expressed. The autonomy of the individual chromosomes in several mitotic movements is shown by their asynchrony during stretch and congression. A considerable elasticity of the chromosomal fiber and of the chromosome body is suggested; and the orientation of a kinetochore is shown to be reversible. The premetaphase stretch is not confined to mantids but is now known to occur in meiosis in phasmids (Hughes-Schrader 1947), and in certain blattids (Matthey 1945, Suomalainen 1946), molluscs (Staiger 1950a and b), and flies (Wolf 1941, 1950). In a less exaggerated form, as already observed in the spermatocytes of the mecopteran Boreus (Cooper 1951), in meiotic prophase of many plant species (Östergren 1951), and in at least one somatic type of mitosis (secondary spermatocytes of the phasmid *Isagoras subaquilus*, (Hughes-Schrader 1947), the pre-metaphase stretch may well prove to be a general feature of the mitotic orientation of chromosomes.

# V. Conclusion

THE PRESENT SURVEY of past attempts to solve the puzzle that is mitosis may well seem disheartening. Not one of the many hypotheses that have been broached has in it the definite promise of a final solution. But we have come to realize that mitosis is comprised of a great complex of different mechanisms. To reach this realization is perhaps a small reward for our efforts, but in order to appreciate it one must remember that such an attitude is not of long standing. The investigations of the past have usually gone out from the assumption that the whole mitotic mechanism when laid bare will prove to be rather simple. In no other way can one explain the fact that nearly all the hypotheses have been built around the idea that a certain, single type of force underlies all mitotic activity, and that variations and adjustments in this force and the cell elements will explain the whole cycle of phenomena with which we are confronted. Thus we have contraction, expansion, hydration, diffusion, and electrical forces, each serving as the core of various hypotheses. The foregoing pages carry the evidence that the solution of the problem cannot be achieved in that way.

The genetic attack has to some extent avoided the complications of multiple effects that other experimental attacks have to deal with all the time. As Bleier (1939) has put it for one type of breeding experiment—"probably there is no more harmless and elegant a method of intrusion into the nucleus"—than the addition of an extra but identical set of chromosomes. This may well be conceded. Nevertheless the breeding experiment is limited in range so far as our problem is concerned. A hybridization, for instance of Orthoptera as Klingstedt (1939) has reported, may upset the synchronization of the spindle apparatus with the chromosomes so that chromatids fail to separate at anaphase. But very similar disturbances may result from X rays, temperature changes, chemicals, mutation, aging, and the presence of supernumeraries

(Bauer 1931, Beadle 1932, Sax 1937, White 1937, Schrader 1941b). It might seem that this would actually give a lead toward an explanation of the effect, but even if a suggestion such as a disturbance of timing relations is correct, no answer has been forthcoming as to how this is done. Similarly Bleier's study (1930a and b) of bivalents and univalents in the meiosis of hybrids indicates a difference in their metaphase maneuvers, but to say as Vandel (1928) did before him that chromosomes must be present in pairs to bring about an orderly mitosis is only one primary step toward a final answer. That genetic mutations may bring about alterations in the mitotic mechanism just as they may in any other feature of the organism requires no affirmation here. Thus certain genes in Zea may bring about extra divisions after meiosis or cause the chromosomes to become sticky (Beadle 1931, 1932); they may affect the polarization of the spindle (Clark 1940, Swanson and Nelson 1942); they may in Drosophila induce a telophase scattering of the chromosomes (Wald 1936); and in Lathyrus disarrange the timing of the spindle processes (Upcott 1937b). But to trace the effect down to the gene does not give us an answer of immediate utility. It tells us that the underlying cause of such a mitotic irregularity is inherited but that does not, in itself, give us an explanation of the chemical and physical reactions that are involved.

The truth is that we badly need more information concerning the physical chemistry of the living cell, and this is now becoming generally recognized. The great increase in the number of investigations dealing with the mitotic problem from that angle is testimony for a gratifying response to that need. It is probably in reaction to this development that during the last ten years there has been a perceptible shift toward a more sophisticated attitude. This is expressed, for instance, in the alterations to which the ancient traction and pushing hypotheses have been subjected and that have gained for them a new respectability.

In the course of the many-sided attack which is now being launched on the mitotic problem, a certain amount of confusion is almost unavoidable. Perhaps it is salutary to remember that there are certain cytological verities to which all conclusions will have to conform. The most obvious of these, as they pertain to the anaphase movements, need hardly be mentioned here. However, there are some equally important facts whose implications are sometimes forgotten in the heat of the battle. They may be summarized as follows:

1. Most of us are now agreed that the fibrous structure of the amphiaster—spindle and asters—primarily represents a special orientation of long particles or micelles. How these fibrous tracts function in the movement of chromosomes is another and unanswered question.

2. More than ever it is realized that the poles or centers play a pivotal role in the mitotic process. But in recent years a second element—the kinetochore—has entered into our considerations as being almost, if not quite, as essential. Any advance in the analysis of either of these two elements will go far toward a final answer to the mitotic question.

3. Whatever the basic mechanism of the anaphase transport of the chromosomes may be, it must not be forgotten that various cellular elements—chromosomes, centers, the nucleus as a whole, and others—are shifted about prior to metaphase without the interposition of a spindle. That should be a sufficient warning that spindle fibers are not the *sine qua non* of all mitotic movements, even though a spindle may be necessary to make such movements orderly and concerted.

The attitude of the past that the answer will some day be revealed to us by a stroke of genius or by luck is almost certainly illusory. The solution will be approached much more surely if we make up our minds that we are confronted by some painstaking work that has to be guided by intelligent planning and a thorough knowledge of the results so far obtained.

# Literature

Except where necessary as a background, no papers published before 1925 are listed here.

Abelson,, P. H., and W. R. Duryee. 1948. Permeability of frog eggs to radioactive ions of a Ringer medium. Anat. Rec., 101: 653–654.

Ahrens, W. 1936. Das dynamische Verhalten der Chromatinschleifen im Stadium des Buketts und das Reduktionsproblem. Zool. Anz., 116: 49–62.

—— 1939. Die Chromosomenbewegungen in der Reifungsphase der Keimzellen. Sitzgber. physik. med. Soz., Erlangen, 70: 73–86.

Alberti, W., and G. Politzer. 1923. Über den Einfluss der Röntgenstrahlen auf die Zellteilung. Arch. mikr. Anat., 100: 83–109.

Andrews, F. M. 1915. Die Wirkung der Zentrifugalkraft auf Pflanzen. Jahrb. wiss. Bot., 56: 221–253.

Astbury, W. T. 1935. The alpha beta transformation of muscle proteins in situ. Nature, 135: 765.

Baltzer, F. 1908. Über mehrpolige Mitosen bei Seeigeleiern. Verh. Phys. Med. Ges., Würzburg, 39: 291–328.

—— 1911. Zur Kenntniss der Mechanik der Kernteilungsfigur. Arch. Entwmk., 32: 500–523.

Banga, I., and A Szent-Györgyi. 1940. Structure proteins. Science, 92: 514.

Barber, H. N. 1939. The rate of movement of chromosomes on the spindle. Chromosoma, 1: 33–50.

Barber, H. N., and H. G. Callan. 1943. The effects of cold and colchicine on mitosis in the newt. Proc. Roy. Soc., London, 131: 258–271.

Bauer, H. 1931. Die Chromosomen von *Tipula paludosa* Meig. in Eibildung und Spermatogenese. Z. Zellf., 14: 138–193.

—— 1938. Die polyploide Natur der Riesenchromosomen. Naturw., 26: 77.

Beadle, G. W. 1931. A gene in maize for supernumerary cell divisions following meiosis. Cornell Univ. Agr. Exp. Sta. Mem., 135: 1–12.

—— 1932. A gene for sticky chromosomes in *Zea mays*. Z. ind. Abst. Vererb., 63: 195–217.

—— 1933. Further studies of asynaptic maize. Cytologia, 4: 269–287.

Beal, J. B. 1932. Microsporogenesis and chromosome behavior in *Nothoscordum bivalve*. Bot. Gaz., 93: 278–295.

Beams, H. W., and R. L. King. 1936. The effect of ultracentrifuging upon chick embryonic cells, with special reference to the "resting" nucleus and the mitotic spindle. Biol. Bul., 71: 188–198.

———— 1938. An experimental study on mitosis in the somatic cells of wheat. Biol. Bull., 75: 189–207.

Beams, H. W., and T. C. Evans. 1940. Some effects of colchicine upon the first cleavage in *Arbacia punctulata*. Biol. Bull., 79: 188–198.

Beck, L. V., and H. Shapiro. 1936. Permeability of germinal vesicle of the starfish egg to water. Proc. Soc. Exp. Biol. Med., 34: 170–172.

Becker, W. A. 1933. Vitalbeobachtungen über den Einfluss von Methylenblau und Neutralrot auf den Verlauf von Karyo- und Zytokinese. Cytologia, 4: 135–157.

———— 1935. Über einige Streitfragen der Zellteilung. Z. Zellf., 23: 253–260.

———— 1936. Vitale Cytoplasma- und Kernfärbungen. Protoplasma, 26: 439–487.

———— 1938. Recent investigations in vivo on the division of the plant cell. Bot. Rev., 4: 446–472.

Belar, K. 1926. Der Formwechsel der Protistenkerne. Jena, Fischer.

Belar, K. 1927. Beiträge zur Kenntniss des Mechanismus der indirekten Kernteilung. Naturw., 15: 725–734.

———— 1928. Die Technik der deskriptiven Cytologie. In Methodik der wissenschaftlichen Biologie: 638–735. Berlin, Springer.

———— 1929a. Beiträge zur Kausalanalyse der Mitose. II. Arch. Entwmk., 118: 359–480.

———— 1929b. Beiträge zur Kausalanalyse der Mitose. III. Z. Zellf., 10: 73–134.

———— 1929c. Review. Collecting Net, 4, August: 5–8.

———— 1930. Über die reversible Entmischung des lebenden Protoplasmas. Protoplasma, 9: 209–244.

Belar, K., and W. Huth. 1933. Zur Teilungsautonomie der Chromosomen. Z. Zellf., 17: 51–66.

Belling, J., 1927. The attachments of chromosomes at the reduction division in flowering plants. J. Genet., 18: 177–205.

———— 1928. Contraction of chromosomes during maturation divisions in Lillium and other plants. Univ. Calif. Publ. Bot., 14: 335–343.

Belling, J., and A. F. Blakeslee. 1924. The distribution of chromosomes in tetraploid Daturas. Am. Nat., 58: 60–70.

Beneden, E. van. 1883. Recherches sur la maturation de l'œuf, la fécondation et la division cellulaire. Arch. Biol., 4: 265–641.
Beneden, E. van, and A. Neyt. 1887. Nouvelles recherches sur la fécondaton et la division mitotique, chez l'Ascaride megalocéphale. Bull. Ac. Roy. Belg., Ser. 3, 14: 215–295.
Benoit, J., and R. Kehl. 1939. Les centromeres au cours de la spermatogenese de la souris. C. r. Soc. Biol., 131: 329–332.
Bensley, R. R. 1938. The gel- and fiber-forming constituent of the protoplasm of the hepatic cell. Anat. Rec., 72: 351.
Bensley, R. R., and N. L. Hoerr. 1934. The chemical basis of the organization of the cell. Anat. Rec., 60: 251.
Berger, C. A. 1938. Multiplication and reduction of somatic chromosome groups as a regular developmental process in the mosquito, *Culex pipiens*. Carneg. Inst. Publ., 496: 211–232.
Bernal, J. D. 1939. A speculation on muscle. In Perspectives in Biochemistry: 45–66. Cambridge, Cambridge Univ. Press.
——— 1940. Structural units in cellular physiology. Publ. Am. Ass. Adv. Sci., 14: 199–205.
Bernal, J. D., and I. Fankuchen. 1941. X-ray and crystallographic studies of plant virus preparations I, II, III. J. Gen. Physiol., 25: 111–165.
Bernstein, J. 1912. Elektrobiologie. Braunschweig, Vieweg.
Berry, R. O. 1941. Chromosome behavior in the germ cells and development of the gonads in *Sciara ocellaris*. J. Morph. 68: 547–583.
Bersa E., and F. Weber. 1922. Reversible Viskositätserhöhung des Cytoplasma unter der Einwirkung des elektrischen Stroms. Ber. deutsch. Bot. Ges., 40: 254–258.
Bertold, G. 1886. Studien über Protoplasmamechanik. Leipzig, Felix.
Bjerknes, V. 1902. Hydrodynamische Fernkräfte. Leipzig, Barth.
——— 1909. Die Kraftfelder. Braunschweig, Vieweg.
Bleie, H. 1930a. Experimentell-zytologische Untersuchungen. I. Z. Zelf., 11: 218–236.
——— 1930b. Untersuchungen über das Verhalten der verschiedenen Komponenten bei der Reduktionsteilung von Bastarden. Cellule, 40: 85-144.
——— 1931. Zur Kausalanalyse der Kernteilung. Genetica, 13: 27–76.
——— 1933. Die Meiosis von Haplodiplonten. Genetica, 15: 129–176.
——— 1934. Bastardkaryologie. Bibl. Genet., 11: 393–489.
——— 1939. Mechanismus der Kernteilung. Arch. exp. Zellf., 22: 25–262.

Bonnevie, K. 1906. Untersuchungen über Keimzellen. Jena. Zeitschr., 41: 229–428.
Botta, B. 1932. Ricerche sulla carica e sul trasporto electtrico della figura cromatica delle cellule in mitosi in culture di cuore embrionale di pollo (Nota preventiva). Arch. exp. Zellf., 12: 455–465.
Boveri, M. 1903. Über Mitosen bei einseitiger Chromosomenbindung. Jena Zeitschur., 37: 401–446.
Boveri, T. 1887. Über die Befruchtung der Eier von *Ascaris megalocephala*. Sitzgber. Ges. Morph. Physiol., München, 3: 71–80.
——— 1888. Zellenstudien. II. Jena, Fischer.
——— 1897. Zur Physiologie der Kern- und Zellteilung. Sitzgber. phys.-med. Ges., Würzburg.
——— 1900. Zellenstudien. IV. Jena, Fischer.
——— 1907. Zellenstudien. VI. Jena, Fischer.
Bresslau, E. 1909. Über die Sichtbarkeit der Zentrosomen in lebenden Zellen. Zool. Anz., 35: 141–145.
Brieger, F. 1934. Ablauf der Meiose bei völliger Asyndese. Ber deutsch. Bot. Ges., 52: 149–153.
Briggs, R., E. U. Green, and T. J. King. 1951. An investigation of the capacity for cleavage and differentiation in *Rana pipiens* eggs lacking functional chromosomes. J. Exp. Zool., 116: 455–500.
Bryan, J. H. D. 1952. DNA-protein relations during microsporogenesis of Tradescantia, Chromosoma, 4: 369–392.
Bucciante, L. 1927. Ulteriori ricerche sulla velocita della mitosi nelle cellule coltivate in funzione della temperatura. Arch. exp. Zellf., 5: 1–24.
Bütschli, O. 1876. Studien über die ersten Entwickelungsvorgänge der Eizelle, die Zellteilungsvorgänge der Eizelle, die Zellteilung und die Konjugation der Infusorien. Abh. Senckenberg. naturf. Ges., 10.
Callan, H. G. 1949. Some physical properties of the nuclear membrane. Exp. Cell Res., 1, Suppl. 1: 48.
——— 1952. A general account of experimental work on amphibian oocyte nuclei. Soc. Exp. Biol., Symp. 6: 243–255.
Callan, H. G., J. T. Randall, and S. G. Tomlin. 1949. An electron microscope study of the nuclear membrane. Nature, 163: 280.
Cannon, H. G. 1923. On the nature of the centrosomal force. J. Genet., 13: 47–78.
Carlson, J. G. 1938a. Mitotic behavior of induced chromosomal fragments lacking spindle attachments in the neuroblasts of the grasshopper. Proc. Nat. Ac. Sci., 24: 500–507.

—— 1938b. Some effects of X-radiation on the neuroblast chromosomes of the grasshopper, *Chortophaga viridifasciata.* Genetics, 23: 596–609.

—— 1946. Protoplasmic viscosity changes in different regions of the grasshopper neuroblast during mitosis. Biol. Bull., 90: 109–121.

—— 1952. Microdissection studies of the dividing neuroblast of the grasshopper, *Chortophaga viridifasciata* (DeGeer). Chromosoma, 5: 200–220.

Carothers, E. E. 1936. Components of the mitotic spindle with special reference to the chromosomal and interzonal fibers in the Acrididae. Biol. Bull., 71: 469–491.

Caspersson, T. 1936. Über den chemischen Aufbau der Strukturen des Zellkerns. Skand. Arch. Physiol., 73, Suppl. 8.

—— 1939. Über die Rolle der Desoxyribosenukleinsäure bei der Zellteilung. Chromosma, 1: 147–156.

—— 1947. The relations between nucleic acid and protein synthesis. Symp. Soc. Exp. Biol., 1: 127–151.

—— 1950. Cell growth and cell function. New York, Norton.

Castro, D. De. 1950. Notes on two cytological problems of the genus Luzula D C. Genet. Iberica, 2: 1–9.

Castro, D. De, A. Camara, and N. Malheiros. 1949. X-rays in the centromere problem of *Luzula purpurea* Link. Genet. Iberica, 1: 1–6.

Chambers, R. 1917. Microdissection studies. II. J. Exp. Zool., 23: 483–505.

—— 1921a. Microdissection studies. III. Biol. Bull., 41: 318–350.

—— 1921b. The formation of the aster in artificial parthenogenesis. J. Gen. Physiol., 4: 33–39.

—— 1924. The physical structure of protoplasm as determined by microdissection and injection. In General Cytology: 237–309. Chicago, Univ. Chicago Press.

—— 1951. Micrurgical studies on the kinetic aspects of cell division. Ann. N.Y. Acad. Sci., 51: 1311–1326.

Chambers, R., and H. B. Fell. 1931. Micro-operations on cells in tissue cultures. Proc. Roy. Soc., B, 109: 380–403.

Chambers, R., and H. C. Sands. 1923. A dissection of the chromosomes in the pollen mother cells of *Tradescantia virginica.* J. Gen. Physiol., 5: 815–819.

Chapochnikov, B. N. 1938. La structure de la figure mitotique dans les spermatocytes de *Potamobious astacus.* Nouvelle méthode de l'étude

des filaments achromatiques. Biol. Zhurnal, 7: 267–273.
Churney, L. 1941. The physico-chemical properties of the nucleus. In Cytology, Genetics and Evolution. Philadelphia, Univ. Pa. Press.
Churney, L., and H. M. Klein. 1937. The electrical charge on nuclear constituents (salivary gland cells of *Sciara coprophila*). Biol. Bull., 72: 384–389.
Clark, F. J. 1940. Cytogenetic studies of divergent meiotic spindle formation in *Zea mays*. Am. J. Bot., 27: 547–559.
Claude, A., and T. S. Potter. 1943. Isolation of chromatin threads from the resting nucleus of leukemic cells. J. Exp. Med., 77: 345–354.
Cleland, R. E. 1926. Meiosis in the pollen mother cells of *Oenothera biennis* and *Oenothera biennis sulfurea*. Genetics, 11: 127–162.
Cleveland, L. R. 1934. The wood-feeding roach Cryptocercus, its protozoa, and the symbiosis between protozoa and roach. Mem. Am. Ac. Arts Sci., 17.
———— 1935a. The centrioles in Pseudotrychonympha and their role in mitosis. Biol. Bull., 69: 46–53.
———— 1935b. The intranuclear achromatic figure of *Oxymonas grandis* sp. nov. Biol. Bull., 69: 54–65.
———— 1935c. The centriole and its role in mitosis as seen in living cells. Science, 81: 598.
———— 1938a. Longitudinal and transverse division in two closely related flagellates. Biol. Bull., 74: 1–41.
———— 1938b. Origin and development of the achromatic figure. Biol. Bull., 74: 41–55.
———— 1938c. Mitosis in Pyrsonympha. Arch. Protistk., 91: 452–455.
———— 1938d. Morphology and mitosis of Teranympha. Arch. Protistk., 91: 442–451.
———— 1949a. The whole life cycle of chromosomes and their coiling systems. Trans. Am. Phil. Soc., 39: 1–100.
———— 1949b. Hormone-induced sexual cycles of flagellates. I. J. Morph., 85: 197–296.
———— 1952. Hormone-induced sexual cycles of flagellates. VIII. J. Morph., 91: 269–323.
Coleman, L. C. 1940. The cytology of *Veltheimia viridifolia* Jacq. Am. J. Bot., 27: 887–895.
Conard, A. 1939. Sur le mécanisme de la division cellulaire et sur les bases morphologiques de la cytologie. Bruxelles, Trav. Jard. Exp.
Conklin, E. G. 1917. Effects of centrifugal force on the structure and development of the eggs of Crepidula. J. Exp. Zool., 22: 311–421.

——— 1931. The development of centrifuged eggs of ascidians. J. Exp. Zool., 60: 1-120.

Cooper, K. W. 1938. Concerning the origin of the polytene chromosomes of Diptera. Proc. Nat. Ac. Sci., 24: 452-458.

——— 1939. The nuclear cytology of the grass mite, *Pediculopsis graminum* (Reut.) with special reference to karymerokinesis. Chromosoma, 1: 51-103.

——— 1941a. Bivalent structure in the fly, *Melophagus ovinus* L. (Pupipara, Hippoboscidae). Proc. Nat. Ac. Sci., 27: 109-114.

——— 1941b. Visibility of the primary spindle fibers and the course of mitosis in the living blastomeres of the mite, *Pediculopsis graminum* Reut. Proc. Nat. Ac. Sci., 27: 480-483.

——— 1944. Analysis of meiotic pairing in Olfersia and considerations of the reciprocal chiasmata hypothesis of sex chromosome conjunction in male Drosophila. Genetics, 29: 537-568.

——— 1945. Normal segregation without chiasmata in female *Drosophila melanogaster*. Genetics, 30: 472-484.

——— 1949. The cytogenetics of meiosis in Drosophila. J. Morph., 84: 81-122.

——— 1951. Compound sex chromosomes with anaphasic precocity in the male mecopteran, *Boreus brumalis* Fitch. J. Morph., 89: 39-57.

Cornman, I. 1944. A summary of evidence in favor of the traction fiber in mitosis. Am. Nat., 78: 410-422.

Cornman, I., and M. E. Cornman. 1951. The action of podophyllin and its fractions on marine eggs. Ann. N.Y. Ac. Sci., 51: 1443-1481.

Creighton, H. B., and B. McClintock. 1931. A correlation of cytological and genetical crossing over in *Zea mays*. Proc. Nat. Ac. Sci., 16: 492-497.

Cretschmar, M. 1928. Das Verhalten der Chromosome bei der Spermatogenese von *Orgia thyellina* Btl. und *antiqua* L. sowie eines ihrer Bastarde. Z. Zellf., 7: 290-309.

Dalcq, A. 1928. Les données expérimentales relatives au mécanisme de la division cellulaire. Biol. Rev., 3: 179-208.

——— 1931. Contribution à l'analyse des fonctions nucléaires dans l'ontogénese de la grenouille. I. Arch. Biol., 41: 143-220.

Dan, K. 1943a. Behavior of the cell surface during cleavage V. J. Fac. Sci. Tokyo Imp. Univ., 6: 297-321.

——— 1943b. On the mechanism of cell division. J. Fac. Sci. Tokyo Imp. Univ., 6: 323.

Darlington, C. D. 1933. Meiosis in Agapanthus and Kniphofia. Cytolo-

gia, 4: 229-240.
——— 1936a. The external mechanics of the chromosomes. Proc. Roy. Soc., London. Ser. B, 121: 264-319.
——— 1936b. Crossing-over and its mechanical relationships in Chorthippus and Stauroderus. J. Genet., 33: 465-500.
——— 1937. Recent advances in cytology. 2d ed. Philadelphia, Blakiston.
——— 1939a. Misdivision and the genetics of the centromere. J. Genet., 37: 341-365.
——— 1939b. The genetical and mechanical properties of the sex chromosomes V. J. Genet., 39: 101-137.
——— 1942. Chromosome chemistry and gene action. Nature, 149: 66-69.
——— 1947. Nucleic acid and the chromosomes. Symp. Soc. Exp. Biol., 1: 252-267.
Darlington, C. D., and P. T. Thomas. 1937. The breakdown of cell division in a Festuca-Lolium derivative. Ann. Bot., 1: 747-762.
Delbrück, M. 1941. A theory of autocatalytic synthesis of polypeptides and its application to the problem of chromosome reproduction. Symp. Quant. Biol., 9: 122-126.
Dembowski, J., and H. Ziegenspek. 1928. Über das Verhalten der Nukleolen bei der Kernteilung in der äussersten Meristemzone von Wurzeln von Helianthus. Bot. Arch., 22: 571-574.
Devisé, R. 1922. La figure achromatique et la plaque cellulaire dans les microsporacytes du *Larix europea*. Cellule, 32: 249-312.
Dobzhansky, T. 1934. Studies on hybrid sterility. I. Z. Zellf., 21: 169-224.
Drüner, L. 1895. Studien über den Mechanismus der Zellteilung. Jena. Zeitschr., 29: 271-344.
Eigsti, O. J. 1940. Effects of colchicine upon the nuclear and cytoplasmic phases of cell division in the pollen tube. Genetics, 25: 116-117.
——— 1947. Colchicine bibliography and supplement by P. Dustin. Lloydia, 10: 65-114.
Eilers, W. 1925. Somatische Kernteilungen bei Coleopteren. Z. Zellf., 2: 593-650.
Ellenhorn, J. 1933. Experimental-photographische Studien der lebenden Zellen. Z. Zellf., 20: 288-308.
Ellerström, S., and J. H. Tjio. 1950. Note on the chromosomes of *Phleum echinatum*. Bot Notis., 4: 443-465.

Elson, D., and E. Chargaff. 1952. On the desoxyribonucleic acid content of sea urchin gametes. Experientia 8: 143-148.

Engelhardt, W. A. 1941. Enzymatic and mechanical properties of muscle proteins. Uspekhi Sovremennoi Biologii, 14: 177-190. (In Russian.)

Ephrussi, B. 1926. Sur les coefficients de température des différentes phases de la mitose des œufs d' oursin (Paracentrotus lividus Lk) et de l'*Ascaris megalocephala*. Protoplasma, 1: 105-123.

—— 1933. Contribution à l'analyse des premiers stades du développement de œuf. Action de la température. Arch. Biol., 44: 1-146.

Fabergé, A. C. 1942. Homologous chromosome pairing: the physical problem. J. Genet., 43: 121-145.

Fankhauser, G. 1929. Über die Beteiligung kernloser Strahlungen (Cytaster) an der Furchung geschnürter Tritoneneier. Rev. Suisse Zool., 36: 179-188.

—— 1934. Cytological studies on egg fragments of the salamander Triton. IV. J. Exp. Zool., 67: 349-395.

Federley, H. 1943. Zytologische Untersuchungen an Mischlingen der Gattung Dicraneura B. (Lepidoptera). Hereditas, 29: 205-254.

—— 1945. Die Konjugation der Chromosomen bei den Lepidopteren. Comment. Biol., Soc. Sci. Fenn., 9: 1-11.

Fischer, A. 1899. Fixierung, Färbung und Bau des Protoplasmas. Jena, Fischer.

Flemming, W. 1879. Zur Kenntniss der Zelle und ihrer Lebenserscheinungen. Arch. mikr. Anat., 16: 302-436.

Fol, H. 1873. Die erste Entwickelung des Geryonideneies. Jena. Zeitschr., 7: 471-492.

Foot, K., and E. C. Strobell. 1905. Prophases and metaphase of the first maturation spindle of *Allolobophora foetida*. Am. J. Anat., 4: 199-243.

Freundlich, H. 1927. Neuere Fortschritte der Kolloidchemie und ihre biologische Bedeutung. Protoplasma, 2: 278-299.

—— 1937. Colloidal structures in biology. J. Phys. Chem., 41: 1151-1161.

Freundlich, H., O. Enslin, and K. Söllner. 1933. Über die Bildung von Taktoiden in Gemischen zweier Sole und ihre biologische Bedeutung. Protoplasma, 17: 489-498.

Frew, P., and R. H. Bowen. 1929. Nucleolar behavior in the mitosis of plant cells. Quart. J. Micr. Sci., 73: 197-214.

Frey-Wyssling, A. 1938. Submikroskopische Morphologie des Protoplasma und seiner Derivate. Berlin, Borntraeger.

——— 1948. Submicroscopic morphology of protoplasm and its derivatives. New York, Elsevier.

Friedrich-Freksa, H. 1940. Bei der Chromosomenkonjugation wirksame Kräfte und ihre Bedeutung für die identische Verdoppelung von Nukleoproteinen. Naturw., 28: 376–379.

Frolowa, S. L. 1936. Struktur der Kerne in den Speicheldrüsen einiger Drosophila-arten. Biol. Zhurnal, 5: 271–292.

Fry, H. J. 1925. Asters in artificial parthenogenesis. I, II. J. Exp. Zool., 43: 11–82.

——— 1928. Conditions determining the origin and behavior of central bodies in cytasters of Echinarachnius eggs. Biol. Bull., 54: 363–395.

——— 1929a. The so-called central bodies in fertilized Echinarachnius eggs. I, II. Biol. Bull., 56: 101–150.

——— 1929b. A critique of the usual concepts concerning the mitotic mechanism of the echinoderm egg. Collecting Net, August: 1–5.

——— 1932. Studies of the mitotic figure. I. Biol. Bull., 63: 149–186.

——— 1933. Studies of the mitotic figure. III. Biol. Bull., 65: 207–237.

——— 1936. Studies of the mitotic figure. V. Biol. Bull., 70: 89–100.

——— 1937. Studies of the mitotic figure. VI. Biol. Bull., 73: 565–590.

Fry, H. J., M. Jacobs, and H. M. Lieb. 1929. The so-called central bodies in fertilized Echinarachnius eggs. III. Biol. Bull., 57: 151–159.

Fry, H. J., and M. E. Parks. 1934. Studies of the mitotic figure. IV. Protoplasma, 21: 473–499.

Fry, H. J., and C. W. Robertson. 1933. Studies of the mitotic figure. II. Anat. Rec., 56: 159–185.

Fujii, K., and K. Yasui. 1936. The structure of the cell and the cell division. Research Hattori-Hokokai Found., 2: 122–127. (Quoted in Shimamura 1940.)

Fürst, E. 1898. Über Centrosomen bei *Ascaris megalocephala*. Arch. milkr. Anat., 52: 97–134.

Geitler, L. 1937. Die Analyse des Kernbaus und der Kernteilung der Wasserläufer *Gerris lateralis* und *Gerris lacustris* und die Somadifferenzierung. Z. Zellf., 26: 641–672.

——— 1938. Chromosomenbau. Berlin, Borntraeger.

Giles, N. H. 1943. The origin of iso-chromosomes at meiosis. Genetics, 28: 512–525.

Goldacre, R. J. 1952. The folding and unfolding of protein molecules as a basis of osmotic work. In Internat. Rev. of Cyt., 1. New York, Academic Press.

Grassé, P. P. 1939. Etudes de mécanique cellulaire: centromeres et centrosomes dans la mitose de certains flagellés. C. r. Soc. Biol., 131: 1015–1018.

Gray, J., 1931. A text book of experimental cytology. New York, Macmillan.

Griffen, A. B., and W. S. Stone, 1940. Studies in the genetics of Drosophila. IX. Univ. Texas Publ., 4032: 201–207.

Gross, F. 1935. Die Reifungs- und Furchungsteilungen von *Artemia salina* in Zusammenhang mit dem Problem des Kernteilungsmechanismus. Z. Zellf., 23: 522–566.

Gurwitsch, A. 1926. Das Problem der Zellteilung physiologisch betrachtet. Monogr. Gesamtg. Pfl. Tiere 11. Berlin, Springer.

Hamburger, H. J. 1904. Osmotischer Druck und Ionenlehre in den medizinischen Wissenschaften. 3. Wiesbaden, J. F. Bergmann.

Harvey, E. B. 1936. Parthenogenetic merogamy or cleavage without nuclei in *Arbacia punctulata*. Biol. Bull., 71: 101–122.

——— 1940. A comparison of the development of nucleate and nonnucleate eggs of *Arbacia punctulata*. Biol. Bull., 79: 166–187.

Heiderich, F. 1910. Sichtbare Centrosomen in überlebenden Zellen. Anat. Anz., 36: 614–618.

Heilbrunn, L. V. 1920. An experimental study of cell-division. I. J. Exp. Zool., 30: 211–239.

——— 1928. The colloid chemistry of protoplasm. Berlin, Borntraeger.

——— 1943. An outline of general physiology. 2d ed. Philadelphia, Saunders.

Heilbrunn, L. V., and K. Daugherty. 1939. The electric charge of protoplasmic colloids. Physiol. Zool., 12: 1–12.

Hinton, T. 1945. A study of the chromosome ends in salivary gland nuclei of Drosophila. Biol. Bull., 88: 144–165.

Hinton, T., and K. C. Atwood. 1941. Terminal adhesions of salivary gland chromosomes in Drosophila. Proc. Nat. Ac. Sci., 27: 491–496.

Hiraoka, T. 1941. Studies of meiosis and mitosis in comparison. Cytologia, 11: 482–492.

Hirayanagi, H. 1929. Chromosome arrangement. III. Mem. Coll. Sci., Kyoto Univ., 4: 272–281.

Hirschler, J. 1935. Über eine Reihe von auf ihre fusomale Natur verdächtiger Zelleinrichtungen. Erg. Fortschr. Zool., 8: 329–414.

——— 1942. Osmiumschwärzung perichromosomaler Membranen in den Spermatocyten der Rhynchoten-Art *Palomena viridissima* Poder. Naturw., 30: 105–106.

Hollande, A. C. 1940. Le role des solenosomes dans les figures de la mitose chez la jacinthe. C. r. Ac. Sci., 210: 342–344.

Howard, A., and S. R. Pelc. 1951. Nuclear incorporation of $P^{32}$ as demonstrated by autoradiographs. Exp. Cell Res., 2: 178–187.

Huettner, A. F. 1924. Maturation and fertilization in *Drosophila melanogaster*. J. Morph., 39: 249–267.

———— 1933. Continuity of the centrioles in *Drosophila melanogaster*. Z. Zellf., 19: 119–134.

Huettner, A. F., and M. Rabinowitz. 1933. Demonstration of the central body in the living cell. Science, 78: 367.

Hughes, A. F., 1952. The mitotic cycle. New York, Academic Press.

Hughes, A. F., and M. M. Swann. 1948. Anaphase movements in the living cell. J. Exp. Biol., 25: 45–70.

Hughes-Schrader, S. 1924. Reproduction in Acroschismus. J. Morph., 39: 157–207.

———— 1931. A study of the chromosome cycle and the meiotic division-figure in *Llaveia bouvari*—a primitive coccid. Z. Zellf., 13: 742–770.

———— 1935. The chromosome cycle of Phenacoccus (Coccidae). Biol. Bull., 69: 462–468.

———— 1940. The meiotic chromosomes of the male *Llaveiella taenechina* Morrison (Coccidae) and the question of the tertiary split. Biol. Bull., 78: 312–338.

———— 1942. The chromosomes of *Nautococcus schraderae* Vays. and the meiotic division figure of male llaveiine coccids. J. Morph., 70: 261–299.

———— 1943a. Meiosis without chiasmata—in diploid and tetraploid spermatocytes of the mantid *Callimantis antillarum* Saussure. J. Morph., 73: 111–141.

———— 1943b. Polarization, kinetochore movements, and bivalent structure in the meiotic chromosomes of male mantids. Biol. Bull., 85: 265–300.

———— 1947. The "pre-metaphase stretch" and kinetochore orientation in phasmids. Chromosoma, 3: 1–21.

———— 1948a. Cytology of coccids (Coccoïdea-Homoptera). Adv. Genet., 2: 127–203.

———— 1948b. Expulsion of the sex chromosome from the spindle in spermatocytes of a mantid. Chromosoma, 3: 257–270.

Hughes-Schrader, S., and H. Ris. 1941. The diffuse spindle attachment of coccids, verified by the mitotic behavior of induced chromosome

fragments. J. Exp. Zool., 87: 429–456.
Huskins, C. L. 1937. The internal structure of chromosomes—a statement of opinion. Cytologia, Fujii Vol.: 1015–1022.
——— 1942. Structural differentiation of nucleus. In the Structure of Protoplasm: 109–126. Ames, Iowa State Coll. Press.
Huskins, C. L., and S. G. Smith, 1935. Meiotic chromosome structure in *Trillium erectum* L. Ann. Bot., 49: 119–150.
Inamdar, N. B. 1949. A note on the reorientation within the spindle of the sex trivalent in a mantid. Biol. Bull., 97: 300–301.
Inoué, S. 1951. A method for measuring small retardations of structures in living cells. Exp. Cell Res., 2: 513–517.
——— 1952a. The effect of colchicine on the microscopic and submicroscopic structure of the mitotic spindle. Exp. Cell Res., Suppl. 2: 305–311.
——— 1952b. Studies on depolarization of light at microscope lens surfaces. I. Exp. Cell Res., 3: 199–208.
——— 1953. Polarization optical studies I. Chromosoma 5: 487–500.
Inoué, S., and K. Dan. 1951. Birefringence of the living cell. J. Morph., 89: 423–456.
Iwata, J. 1940. Studies on chromosome structure. II. Jap. J. Bot., 10: 375–382
Jacobson, W., and M. Webb. 1951. The two types of nucleic acid during mitosis. J. Physiol., 112: 2p–4p.
——— 1952. The two types of nucleoproteins during mitosis. Exp. Cell Res., 3: 163–183.
Jehle, H. 1950. Quantum-mechanical resonance between identical molecules. J. Chem. Phys., 18: 1150–1164.
Johnson, H. H. 1931. Centrioles and other cytoplasmic components of the male germ cells of the Gryllidae. Z. wiss. Zool., 140: 115–166.
Jollos, V., and T. Peterfi. 1923. Furchung von Axolotleiern ohne Beteiligung des Kernes. Biol. Centralbl., 43: 286–291.
Juel, H. O. 1897. Die Kerntheilungen in den Pollenmutterzellen von *Hemerocallis fulva* und die bei denselben auftretenden Unregelmässigkeiten. Jahrb. wiss. Bot., 30: 205–226.
Jungers, V. 1931. Figures caryocinétiques et cloisonnement du protoplasme dans l'endosperm d'*Iris pseudo-acorus*. Cellule, 40: 5–82.
——— 1934. Mitochondries, chromosomes et fuseau dans les sporocytes d'*Equisetum limosum*. Cellule, 43: 321–340.
Kamiya, N. 1937. Untersuchungen über die Wirkung des elektrischen Stromes auf lebende Zellen. Cytologia, Fujii Vol.: 1036–1042.

Kattermann, G. 1939. Ein neuer Karyotyp bei Roggen. Chromosoma, I: 284-299.

Kaufmann, B. P. 1934. Somatic mitoses of *Drosophila melanogaster*. J. Morph., 56: 125-155.

―――― 1948. Chromosome structure in relation to the chromosome cycle. Bot. Rev., 14: 57-126.

Kaufmann, B. P., M. R. McDonald, and H. V. Gay. 1951. The distribution and interrelation of nucleic acids in fixed cells as shown by enzymatic hydrolysis. J. Cell. Comp. Physiol., 38: 71-100.

Klein, E. 1878. Observations on the structure of cells and nuclei. Quart. J. Micr. Sci., 18: 315-339.

―――― 1879. Observations on the glandular epithelium and division of nuclei in the skin of Newt. Quart. J. Micr. Sci., 19: 404-421.

Klingstedt, H. 1939. Taxonomic and cytological studies on grasshopper hybrids. I. J. Genet., 37: 389-421.

Koerperich, J. 1930. Étude comparative du noyau, chromosomes et de leurs relations avec le cytoplasme. Cellule, 39: 306-401.

Koller, P. C. 1934. The movements of the chromosomes within the cell and their dynamic interpretation. Genetica, 16: 447-467.

―――― 1938. Asynapsis in *Pisum sativum*. J. Genet., 36: 275-306.

―――― 1939. A new race of *Drosophila miranda*. J. Genet., 38: 477-493.

Koonz, C. H. 1936. Some unusual cytological phenomena in the spermatogenesis of a haploid parthenogenetic hymenopteran, *Aenoplex smithii* (Packard). Biol. Bull., 71: 375-385.

Koslov, V. E. 1937. Observations on the kinetochore of mitotic chromosomes. Biol. Zhurnal, 6: 759-767.

Kostoff, D. 1938. The effect of centrifuging upon the germinated seeds. Cytologia, 8: 420-442.

Kostoff, D., and N. Arutinian. 1938. Heterochromatic (inert) regions in the chromosomes of *Crepis capillaris*. Nature 141: 514-515.

Kupka, E. 1950. Die Mitosen- und Chromosomenverhältnisse bei der grossen Schwebrenke, *Coregonus warturanni* (Bloch) des Attersees. Öster. Zool. Z., 2: 605-623.

Kupka, E., and F. Seelich. 1948. Die anaphasische Chromosomenbewegung. Chromosoma, 3: 302-327.

Küster, E. 1935. Die Pflanzenzelle. Jena, G. Fischer.

―――― 1948. Über Eiweissspindeln von Impatiens. Biol. Centralbl., 67: 27-31.

Kuwada, Y. 1929. Chromosome arrangement. I. Mem. Coll. Sci., Kyoto

Univ., 4: 199–264.
Lagerstedt, S. 1949. Cytological studies on the protein metabolism of the liver in rat. Acta Anat., Suppl. 9: 3–116.
Lamb, A. B. 1907. A new explanation of the mechanics of mitosis. J. Exp. Zool., 5: 27–33.
Lams, H. 1910. Recherches sur l'œuf d'*Arion empiricorum* (Fér). Acad. roy. Belg., Classe de Sciences, 2: 1–170.
Landau, E. 1910. Einige Worte zur karyokinetischen Zellteilung. Biol. Centralbl., 30: 646–650.
Lawrence, W. J. C. 1931. The secondary association of chromosomes. Cytologia, 2: 352–384.
Lehotzky, P. von. 1935. Die Wirkung des elektrischen Stromes auf lebende Zellen und Gewebe. Arch. exp. Zellf., 18: 3–12.
Lenoir, M. 1932. Évolution vivante d'une anaphase I (= hétérotypique) et d'une télophase II (= homéotypique) dans les cellules-mères du grain de pollen chez le *Lilium candidum* L. Rev. gen Bot., 44: 140–144.
Lettré, H. 1948. Mitosegifte und cancerogene Faktoren als Antibiotica. Z. Krebsf., 56: 5–35.
Lettré, H., and R. Lettré. 1947. Aufhebung der Mitosegiftwirkung metallorganischer Verbindungen. Naturwiss., 34: 127.
Leuchtenberger, C., G. Klein, and E. Klein. 1952. The estimation of nucleic acids in isolated nuclei of ascites tumors by ultraviolet microspectrophotometry and its comparison with the chemical analysis. Cancer Res., 12: 480–483.
Leuchtenberger, C., R. Leuchtenberger, R. Vendrely, and C. Vendrely. 1952. The quantitative estimation of desoxyribose nucleic acid (DNA) in isolated individual animal nuclei by the Caspersson ultraviolet method. Exp. Cell Res., 3: 240–244.
Leuchtenberger, C., and H. Z. Lund. 1952. A cytochemical study of desoxyribose nucleic acid in senile keratosis. Cancer Res., 12: 278.
Levan, A. 1941. Syncyte formation in the pollen mother-cells of haploid *Phelum pratense*. Hereditas, 27: 244–253.
Levine, M. 1951. The action of colchicine on cell division in human cancer, animal and plant tissues. Ann. N.Y. Ac. Sci., 51: 1365–1408.
Levitsky, G. A. 1931. The morphology of chromosomes. Bull. Appl. Bot., 27: 19–173.
Lewis, M. R. 1923. Reversible gelation in living cells. Johns Hopkins Hosp. Bull., 34: 373–379.
——— 1934. Reversible solation of the mitotic spindle of living chick

embryo cells studied in vivo. Arch. exp. Zellf., 16: 159–167.
Lewis, W. H., and M. R. Lewis. 1917. The duration of the various phases of mitosis in the mesenchyme cells of tissue cultures. Anat. Rec., 13: 349–367.
———— 1924. Behavior of cells in tissue cultures. In General Cytology: 383–449. Chicago, Univ. Chicago Press.
Lillie, F. R. 1906. Observations and experiments concerning the elementary phenomena of development in Chaetopterus. J. Exp. Zool., 3: 153–269.
———— 1909. Karyokinetic figures of centrifuged eggs; an experimental test of the center of force hypothesis. Biol. Bull., 17: 101–120.
Lillie, R. S. 1903. On differences in the direction of the electrical convection of certain free cells and nuclei. Am. J. Physiol., 8: 273–283.
———— 1905a. On the conditions determining the disposition of the chromatic filaments and chromosomes in mitosis. Biol. Bull., 8: 193–204.
———— 1905b. The physiology of cell division. I. Am. J. Physiol., 15: 46–84.
———— 1909. The general biological significance of changes in the permeability of the surface layer or plasma-membrane of living cells. Biol. Bull., 17: 188–208.
———— 1910. The physiology of cell division. II. Am. J. Physiol. 26: 106–133.
———— 1911. The physiology of cell division. IV. J. Morph. 22: 695–730.
———— 1913. The physiology of cell division. V. J. Exp. Zool., 15: 23–47.
———— 1916. The physiology of cell division. VI. J. Exp. Zool., 21: 369–402.
Lima-De-Faria, A. 1949a. The structure of the centromere of the chromosomes of rye. Hereditas, 35: 77–85.
———— 1949b. Genetics, origin and evolution of kinetochores. Hereditas, 35: 422–444.
———— 1950. The Feulgen test applied to centromeric chromomeres. Hereditas, 36: 60–74.
Lison, L., and N. Fautrez-Firlefyn. 1950. Desoxyribonucleic acid content of ovarian cells in *Artemia salina*. Nature, 166: 610–611.
Lison, L., and J. Pasteels. 1950. Mesures photométriques de la teneur en acide désoxyribosenucléique des noyaux au cours de la mitose. Bull. Ac. Roy. Belgique, 36: 348–354.

Lorbeer, G. 1934. Die Zytologie der Lebermoose mit besonderer Berücksichtigung allgemeiner Chromosomenfragen. Jahrb. wiss. Bot., 80: 567–818.

Love, R. M. 1943. A cytogenetic study of off types in a winter wheat, Dawson's Golden Chaff, including a white chaff mutant. Canad. J. Res., 21: 257–264.

Lucas, F. F., and M. B. Stark. 1931. A study of living sperm cells of certain grasshoppers by means of the ultraviolet microscope. J. Morph., 52: 91–115.

Lundegardh, H. 1912. Chromosomen, Nukleolen und die Veränderungen im Protoplasma bei der Karyokinese. Beitr. Biol. Pflanz., 11: 273–542.

Luyet, B. J. 1935. Behavior of the spindle fibers in centrifuged cells. Proc. Soc. Exp. Biol. Med., 33: 163–165.

Luyet, B. J., and R. A. Ernst. 1934. Some physical properties of the nuclear membrane. Proc. Soc. Exp. Biol. Med., 31: 1225–1227.

McAllister, F. 1931. The formation of the achromatic figure in *Spirogyra setiformis*. Am. J. Bot., 18: 838–853.

McClendon, J. F. 1908. The segmentation of eggs of *Asterias forbesii* deprived of chromatin. Arch Entwmk., 26: 662–668.

——— 1913. The laws of surface tension and their applicability to living cells and cell division. Arch. Entwmk., 37: 233–247.

McClintock, B. 1932. A correlation of ring-shaped chromosomes with variegation in *Zea mays*. Proc. Nat. Ac. Sci., 18: 677–681.

——— 1933. The association of non-homologous parts of chromosomes in the mid-prophase of mieosis in *Zea mays*. Z. Zellf., 19: 191–237.

——— 1938. The fusion of broken ends of sister half-chromatids following chromatid breakage at meiotic anaphasees. Univ. Missouri Coll. Agr. Bull. 290.

———1939. The behavior in successive nuclear divisions of a chromosome broken at meiosis. Proc. Nat. Ac. Sci., 25: 405–416.

——— 1942. The fusion of broken ends of chromosomes following nuclear fusion. Proc. Nat. Ac. Sci., 28: 458–463.

Maeda, T., and K. Kato. 1929. Chromosome arrangement. VII. Mem. Coll. Sci., Kyoto Univ., 4: 327–345.

Malheiros, N., D. De Castro, and A Camara. 1947. Chromosomas sem centrómero localizado. O caso de *Luzula purpurea* Link. Agronomia Lusitana, 9: 51–71.

Martens, P. 1927. Recherches expérimentales sur la cinèse dans la cellule vivante. Cellule, 38: 67–174.

Martens, P. 1928. Les structures nucléaires et chromosomiques dans la cellule vivante et dans la cellule fixée. Bull. Hist. appl., 5: 229-252.
―――― 1929. Nouvelles recherches expérimentales sur la cinèse dans la cellule vivante. Cellule, 39: 169-216.
Matsuura, H. 1937. Chromosome studies on *Trillium kamtschaticum* Pall. III. Cytologia, 8: 142-177.
―――― 1941. Chromosome studies in *Trillium kamtschaticum* Pall. XIII. Cytologia, 11: 369-379.
―――― 1949a. Chromosome studies on *Trillium kamtschaticum* Pall. and its allies XVIII. Chromosoma, 3: 418-430.
―――― 1949b. Chromosome studies on *Trillium kamtschaticum* Pall. and its allies XXII. Chromosoma, 3: 431-439.
Matsuura, H., and T. Haga. 1942. Chromosome studies on *Trillium kamtschaticum* Pall. Cytologia, 12: 397-417.
Matsuura, H., and M. Kurabayashi. 1951. Chromosome studies on *Trillium kamtschaticum* Pall. and its allies XXIV. Chromosoma 4: 273-283.
Matthey, R. 1945. Cytologie de la parthénogénèse chez *Pycnoscelus surinamensis* L. (Blattariae-Blaberidae-Panchlorinae). Rev. Suisse Zool., 52, Suppl. 1: 1-109.
Matthey, R., and J. Aubert. 1947. Les chromosomes des plécoptères. Bull. Biol., 81: 202-246.
Mazia, D. 1941. Enzyme studies on chromosomes. Cold Spring Harbor Symp., 9: 40-46.
―――― 1949. Desoxyribonucleic acid and desoxyribonuclease in development. Growth Symposium 9: 5-31.
Mazia, D., and L. Jaeger, 1939. Nuclease action, protease action and histochemical tests on salivary chromosomes of Drosophila. Proc. Nat. Ac. Sci., 25: 456-461.
Mazia, D., and K. Dan. 1952. The isolation and biochemical characterization of the mitotic apparatus of dividing cells. Proc. Nat. Ac. Sci., 38: 826-838.
Melander, Y. 1950 Studies on the chromosomes of *Ulophysema öresundense*. Hereditas, 36: 233-255.
Metz, C. W. 1933. Monocentric mitosis with segregation of chromosomes in Sciara and its bearing on the mechanism of mitosis. Biol. Bull., 64: 333-347.
―――― 1934. The role of the "chromosome sheath" in mitosis and its possible relation to phenomena of maturation. Proc. Nat. Ac. Sci., 20: 159-163.
―――― 1936. Factors influencing chromosome movements in mitosis. Cytologia, 7: 219-231.

Metz, C. W., M. S. Moses, and E. N. Hoppe. 1926. Chromosome behavior and genetic behavior in Sciara (Diptera). Z. ind. Abst. Vererb., 42: 237–270.

Metz, C. W., and J. F. Nonidez. 1924. The behavior of the nucleus and chromosomes during spermatogenesis in the robber fly *Lasiopogon bivittaus*. Biol. Bull., 46: 153–167.

Metzner, R. 1894. Beiträge zur Granulalehre. I. Arch. Anat. Physiol., 309–348.

Meves, F. 1897a. Über die Entwickelung der männlichen Geschlechtszellen von *Salamandra maculosa*. Arch. mikr. Anat., 48: 1–83.

—— 1897b. Zellteilung. Ergeb. Anat., 6: 284–390.

—— 1897c. Über Centralkörper in männlichen Geschlechtszellen von Schmetterlingen. Anat. Anz., 14: 1–6.

Meyer, K. H. 1928. Über den Aufbau des Seiden-Fibroins. Ber. deutsch. chem. Ges., 61: 1932–1936.

—— 1929. Über Feinbau, Festigkeit und Kontraktilität tierischer Gewebe. Biochem. Z., 214: 253–281.

Meyer, K. H., and H. Mark. 1932. Zur Cellulose-Frage. Bemerkungen zu einer gleichgenannten Arbeit von Kurt Hesse und Carl Trogus. Ber. deutsch. chem. Ges., 61: 2432–2436.

Michel, K. 1941. Die Darstellung von Chromosomen mittels des Phasenkontrastverfahrens. Naturwiss., 29: 61–62.

—— 1943. Die Kern- und Zellteilung im Zeitrafferfilm. Zeiss-Nachrichten, 4: 236–251.

Minouchi, O. 1929. Chromosome arrangement. VI. Mem. Coll. Sci., Kyoto Univ., 4: 323–326.

Minouchi, O., and S. Iriki. 1931. On the chromosomes of *Bufo bufo japonicus* Schlegel. Mem. Coll. Sci., Kyoto Univ., 6: 39–52.

Mirsky, A. E. 1936. Protein coagulation as a result of fertilization. Science, 84: 333.

Mirsky, A. E., and H. Ris. 1947a. Isolated chromosomes. J. Gen. Physiol., 31: 1–6.

—— 1947b. The chemical composition of isolated chromosomes. J. Gen. Physiol., 31: 7–18.

Mohr, O. L. 1919. Mikroskopische Untersuchungen zu Experimenten über den Einfluss der Radiumstrahlen und der Kältewirkung auf die Chromatinreifung und das Heterochromosom bei *Decticus verrucævorus*. Arch. mikr. Anat., 92: 300–368.

Möllendorf, W. von. 1937a. Zur Kenntniss der Mitose. II. Z. Zellf., 27: 301–325.

—— 1937b. Beiträge zum Problem der Zellenviscosität. Arch. exp.

Zellf., 19: 263-275.
— 1938a. Zur Kenntniss der Mitose. I. Arch. exp. Zellf., 21: 1-66.
— 1938b. Zur Kenntniss der Mitose. IV. Z. Zellf., 28: 512-546.
— 1939a. Zur Kenntniss der Mitose. VIII. Z. Zellf., 29: 706-749.
— 1939b. Durch carcinogene Kohlenwasserstoffe und Geschlechtshormone in Gewebekulturen erzielte Mitosestörungen. Klin. Wschr.: 1098-1099.
Möllendorf, W. von, and G. Laqueur. 1938. Zur Kenntniss der Mitose. III. Z. Zellf., 28: 319-340.
Möllendorf, W. von, and M. Ostrouch. 1939. Zur Kenntniss der Mitose. VII. Z. Zellf., 29: 323-355.
Monné, L. 1935. Permeability of the nuclear membrane to vital stains. Proc. Soc. Exp. Biol. Med., 32: 1197-1198.
Morgan, T. H. 1910. Cytological studies of centrifuged eggs. J. Exp. Zool., 9: 593-657.
Morgan, T. H., and A. Tyler. 1935. Effects of centrifuging eggs of Urechis before and after fertilization. J. Exp. Zool., 70: 301-341.
Mottier, D. M. 1897. Beiträge zur Kenntniss der Kerntheilung in den Pollenmutterzellen einiger Dikotylen und Monokotylen. Jahrb. wiss. Bot., 30: 169-204.
— 1903. The behavior of the chromosomes in the spore mother cells of higher plants and the homology of the pollen and embryo-sac mother cells. Bot. Gaz., 35: 250-280.
Muller, H. J. 1941. Induced mutations in Drosophila. Symp. Quant. Biol., 9: 151-165.
Muto, A. 1929. Chromosome arrangement. II. Mem. Coll. Sci., Kyoto Univ., 4: 265-271.
Nagao, S. 1929. Chromosome arrangement. VIII. Mem. Coll. Sci., Kyoto Univ., 4: 346-352.
Nakamura, K. 1931. Studies on reptilian chromosomes. II. Cytologia, 2: 385-402.
Nakamura, T. 1929. Chromosome arrangement. IX. Mem. Coll. Sci., Kyoto Univ., 4: 353-369.
Naville, A., and J. de Beaumont. 1933. Recherches sur les chromosomes des neuropteres. Arch. Anat. micr., 29: 199-243.
Nawaschin, S. 1912. Über den Dimorphismus der Kerne in den somatischen Zellen bei *Galtonia candicans* (in Russian). Bull. Ac. Imp. Sci., Petersburg, 22: 373-386.
— 1927. Zellendimorphismus bei *Galtonia candicans* Des. und einigen verwandten Monokotylen. Ber. deutsch. Bot. Ges., 45: 415-429.

Nebel, B. R. 1935. Chromosomenstruktur. VI. Züchter, 7: 132–136.
— 1939. Chromosome structure. Bot. Rev., 5: 563–627.
Nemec, B. 1927. Über die Beschaffenheit der achromatischen Teilungsfigur. Arch. exp. Zellf., 5: 77–82.
— 1929. The mechanism of mitotic division. Proc. Intern. Cong. Plant Sci., 1: 243–249.
Nishiyama, I. 1931. The genetics and cytology of certain cereals. Jap. J. Genet., 7: 49–102.
— 1933. The genetics and cytology of certain cereals. IV. Jap. J. Genet., 8: 107–124.
Ogawa, K. 1929. Chromosome arrangement. V. Mem. Coll. Sci., Kyoto Univ., 4: 309–322.
Ogur, M., R. O. Erickson, G. U. Rosen, K. B. Sax, and C. Holden. 1951. Nucleic acids in relation to cell division in *Lilium longifolium*. Exp. Cell Res., 2: 73–89.
Oksala, T. 1943. Zytologische Studien an Odonaten I. Ann. Ac. Sci. Fenn., 4: 1–63.
— 1944. Zytologische Studien an Odonaten II. Ann. Ac. Sci. Fenn., 4: 1–33.
— 1945. Zytologische Studien an Odonaten III. Ann. Ac. Sci. Fenn., 4: 1–29.
Östergren, G. 1943. Elastic chromosome repulsion. Hereditas, 29: 444–450.
— 1944a. An efficient chemical for the induction of sticky chromosomes. Hereditas, 30: 213–216.
— 1944b. Colchicine mitosis, chromosome contraction, narcosis, and protein chain folding. Hereditas, 30: 429–467.
— 1945a. Parasitic nature of extra fragment chromosomes. Bot. Notis., 1: 157–163.
— 1945b. Transverse equilibrium on the spindle. Bot. Notis., 1: 467–468.
— 1945c. Equilibrium of trivalents and the mechanism of chromosome movements. Hereditas, 31: 498.
— 1947a. Proximal heterochromatin, structure of the centromere and the mechanism of its misdivision. Bot. Notis., 2: 176–177.
— 1947b. Heterochromatic B-chromosomes in Anthoxanthum. Hereditas, 33: 261–296.
— 1948a. Chromosome bridges and breaks by cosmarin. Bot. Notis., 2: 376–380.
— 1948b. Chromatin stains of Feulgen type involving other dyes than fuchsin. Hereditas, 34: 510–511.

Östergren, G. 1949a. Luzula and the mechanism of chromosome movements. Hereditas, 35: 445–468.
—— 1949b. A survey of factors working at mitosis. Hereditas, 35: 525–528.
—— 1949c. Equilibria and movements of chromosomes. Proc. Eighth Int. Cong. Genetics: 688–689.
—— 1950a. Considerations on some elementary features of mitosis Hereditas, 36: 1–18.
—— 1950b. Cytological standards for the quantitative estimation of spindle substances. Hereditas, 36: 371–382.
—— 1950c. Isopycnosis and isopycnotic, two new terms for use in chromosome studies. Hereditas, 36: 511–513.
—— 1951. The mechanism of co-orientation in bivalents and multivalents. Hereditas, 37: 85–156.
Östergren, G., and R. Prakken. 1946. Behavior on the spindle of the actively mobile chromosome ends of rye. Hereditas, 32: 473–494.
Osterhout, W. J. V. 1897. Über Entstehung der karyokinetischen Spindel bei Equisetum. Jahrb. wiss. Bot., 30: 159–169.
Painter, T. S. 1916. Contributions to the study of cell mechanics. I. J. Exp. Zool., 20: 509–529.
—— 1918. Contributions to the study of cell mechanics. II. J. Exp. Zool., 24: 445–498.
Painter, T. S., and E. Reindorp. 1939. Endomitosis in the nurse cells of the ovary of *Drosophila melanogaster*. Chromosoma, 1: 267–283.
Painter, T. S., and W. Stone. 1935. Chromosome fusion and speciation in Drosophila. Genetics, 20: 327–342.
Pasteels, J., and L. Lison. 1950a. Teneur des noyaux au repos en acide désoxyribonucléique dans differents tissus chez le rat. Compt. rend. Acad. Sci., 230: 780–782.
—— 1950b. Recherches histophotométriques sur le teneur en acide désoxyribosenucléique au cours de mitoses somatiques. Arch. Biol., 61: 445–474.
Payne, F. 1909. Some new types of chromosome distribution and their relation to sex. Biol. Bull., 16: 119–166.
Pease, D. C. 1941. Hydrostatic pressure effects upon the spindle figure and chromosome movement. J. Morph., 69: 405–441.
—— 1946. Hydrostatic pressure effects upon the spindle and chromosome movement II. Biol. Bull., 91: 145–169.
Pfeiffer, H. H. 1939. Experimentelle Beiträge zur Mitosephysik. Arch. exp. Zellf., 22: 263–267.

——— 1940. Experimentelle Cytologie. Leyden, Chronica Botanica Co.
Pilawski, S. 1933a. Le corps mitochondrial dans la spermatogenèse chez *Cicindela hybrida* L. C. r. Soc. Biol., 113: 189–192.
——— 1933b. Les structures fibrillaires osmiophiles dans la spermatogenèse chez *Cicindela hybrida* L. (Coleoptera). C. r. Soc. Biol., 114: 346–348.
Pinney, E. 1908. Organization of the chromosomes in *Phrynotettix magnus*. Kansas Univ. Sci. Bull., 4: 309–316.
Piza, S. de T. 1939a. Comportamento dos cromossomios na primeira divisao do espermatocito de *Tityus bahiensis*. Sci. Genet., 1: 255–261.
——— 1939b. Consideracões em torno da meiose do *Tityus bahiensis* (Scorpiones-Buthidae) e una nova teoria sobre a movimentacão dos cromossomios. Jorn. Agron., 2: 343–370.
——— 1941. Chromosomes with two spindle attachments. J. Hered., 32: 423–426.
——— 1943. The uselessness of the spindle fibers for moving the chromosomes. Am. Nat., 77: 442–462.
——— 1950. The present status of the question of the kinetochore. Genet. Iberica, 2: 193–199.
Politzer, G. 1934. Pathologie der Mitose. Berlin, Borntraeger.
Pollister, A. W. 1933. Notes on the centrioles of amphibian tissue cells. Biol. Bull., 65: 529–545.
——— 1939. Centrioles and chromosomes in the atypical spermatogenesis of Vivipara. Proc. Nat. Ac. Sci., 25: 189–195.
——— 1941. Mitochondrial orientations and molecular patterns. Physiol. Zool., 14: 268–280.
——— 1952. Nucleoproteins of the cell nucleus. Exp. Cell Res., Suppl. 2: 59–74.
Pollister, A. W., and A. E. Mirsky. 1943. The isolation of chromosomes from resting nuclei. Genetics, 28: 86.
Pollister, A. W., and P. F. Pollister. 1943. The relation between centriole and centromere in atypical spermatogenesis of viviparid snails. Ann. N.Y. Ac. Sci., 45: 1–48.
Pollister, A. W., and H. Ris. 1947. Nucleoprotein determinations in cytological preparations. Cold Spring Harbor Symp., 12: 147–157.
Pollister, A. W., H. Swift, and M. Alfert. 1951. Studies on the desoxypentose nucleic acid content of animal nuclei. J. Cell. Comp. Physiol., 38: 101–120.
Prakken, R., and A. Müntzing. 1942. A meiotic peculiarity in rye, simulating a terminal kinetochore. Hereditas, 28: 441–482.

Propach, H. 1940. Die Centromeren in der Pollenkornmitose von *Tradescantia gigantea* Rose. Chromosoma, 1: 521-526.

Rabinowitz, M. 1941. Studies on the cytology and early embryology of the egg of *Drosophila melanogaster*. J. Morph., 69: 1-50.

Rashevsky, N. 1938. Mathematical biophysics: Physicomathematical foundations of biology. Chicago, Univ. Chicago Press.

—— 1940. Advanced applications of mathematical biology. Chicago, Univ. Chicago Press.

—— 1941. Some remarks on the movement of chromosomes during cell division. Bull. Math. Biophys., 3: 1-3.

—— 1948. Mathematical biophysics. Rev. ed., Chicago, Univ. Chicago Press.

Regemorter, D. van. 1926. Les troubles cinétiques dans les racines chloralisées et leur portée pour l'interprétation des phénomènes normaux. Cellule, 37: 43-73.

Reinke, F. 1900. Zum Beweis der trajektoriellen Natur der Plasmastrahlungen. Arch. Entwmk., 9: 410-424.

Reiss, P. 1947. Interpretation physico-chemique du mécanisme de la mitose. Actualités Scientifiques et Industrielles, 1033: 1-39.

Rhoades, M. M. 1940. Studies of a telocentric chromosome in maize with reference to the stability of its centromere. Genetics, 25: 483-521.

Rhoades, M. M., and W. E. Kerr. 1949. A note on centromere organization. Proc. Nat. Ac. Sci., 35: 129-132.

Rhoades, M. M., and H. Vilkomerson. 1942. On the anaphase movement of chromosomes. Proc. Nat. Ac. Sci., 28: 433-436.

Ribbands, C. R. 1941. Meiosis in Diptera. I. J. Genet., 41: 411-442.

Ries, E. 1938. Grundriss der Histophysiologie; Allgemeine Methoden und Probleme. Leipzig, Akad. Verlagsg.

Ries, E., and M. Gersch. 1936. Gibt das fixierte Präparat ein Äquivalentbild der lebenden Zelle? Z. Zellf., 25: 14-33.

Ris, H. 1942. A cytological and experimental analysis of the meiotic behavior of the univalent X chromosome in the bearberry aphid *Tamalia* ( = Phyllaphis) *coweni* (Ckll). J. Exp. Zool. 90: 267-330.

—— 1943. A quantitative study of anaphase movement in the aphid Tamalia. Biol. Bull., 85: 164-179.

—— 1949. The anaphase movement of chromosomes in the spermatocytes of the grasshopper. Biol. Bull., 96: 90-106.

Robertson, W. R. B. 1930. Chromosome studies V. J. Morph. 50: 209-258.

Robyns, W. 1924. La fuseau de caryocinèse et le fuseau de cytocinèse. Cellule, 34: 367-454.

―――― 1929. La figure achromatique, sur materiel frais, dans les divisions somatiques des phanerogames. Cellule, 39: 83-121.

Rosen, F. 1925. Zur Mechanik der indirekten Kernteilung. Ber. deutsch. Bot. Ges., 43: 211-217.

Runnström, J. 1936. Über die Veränderung der Plasmakolloide bei der Entwicklungserregung des Seeigels. II. Protoplasma, 5: 201-310.

Saez, F. A. 1941. Alteraciones experimentales inducidas por la acción de la gravedad en las células somáticas de *Lathyrus odoratus* (Leguminosae). An. Soc. Cient. Argentina, 132: 139-150.

Sakamura, T. 1920. Experimentelle Studien über die Zell- und Kernteilung. J. Coll. Sci., Univ. Tokyo, 39: 1-221.

Salazar, A. L. 1925. Sur l'existence de mouvements internes dans la sphère attractive au repos. C. r. Ass. Anat., 20. Réunion Paris.

Sax, K. 1937. Effect of variations in temperature on nuclear and cell division in Tradescantia. Am. J. Bot., 24: 218-255.

Sax, K., and J. G. O'Mara. 1941. Mechanism of mitosis in pollen tubes. Bot. Gaz., 102: 629-636.

Schaede, R. 1925a. Untersuchungen über Zelle, Kern und ihre Teilung am lebenden Objekt. Beitr. Biol. Pflanz., 14: 231-261.

―――― 1925b. Über den Bau der Spindelfigur. Beitr. Biol. Pflanz., 14: 367-385.

―――― 1927. Vergleichende Untersuchungen über Cytoplasma, Kern und Kernteilung im lebenden und fixierten Zustand. Protoplasma, 3: 145-191.

―――― 1928. Über das Verhalten der Nukleolen während der Kernteilung. Protoplasma, 5: 41-54.

―――― 1929. Kritische Untersuchungen über die Mechanik der Karyokinese. Planta, 8: 383-397.

―――― 1930. Zentrifugalversuche mit Kernteilungen. Planta, 11: 243-262.

―――― 1931. Über einige Probleme der Kernteilung. Beitr. Biol. Pflanz., 19: 141-178.

―――― 1936. Untersuchungen mit der Nuklealreaktion an Kern und Kernteilung. Planta, 26: 167-192.

―――― 1937. Anordnung und Gestalt der Chromosomen von *Galtonia candicans*. Ber. deutsch. Bot. Ges., 55: 485-492.

Schmidt, W. J. 1936a. Kernspindel und Chromosomen im lebenden sich furchenden Ei von *Psammechinus miliaris* (Müll). Ber. oberhess.

Ges. Nat. Heilk., 17: 140.

―――― 1936b. Doppelbrechung von Chromosomen und Kernspindel in der lebenden Zelle. Naturw., 24: 463.

―――― 1936c. Die Doppelbrechung der Kernspindel und ihre Feinstruktur, ein Argument für die Zugfasertheorie und die Fadenmolikeltheorie. Biodynamica, 22.

―――― 1937a. Die Doppelbrechung von Karyoplasma, Zytoplasma, und Metaplasma. Protopl. Monogr., 11. Berlin, Borntraeger.

―――― 1937b. Die Doppelbrechung von Chromosomen und Kernspindel und ihre Bedeutung für das kausale Verständniss der Mitose. Arch. exp. Zellf., 19: 352-360.

―――― 1939. Doppelbrechung der Kernspindel und Zugfasertheorie der Chromosomenbewegung. Chromosoma, 1: 253-264.

―――― 1940. Neuere polarisationsoptische Arbeiten auf dem Gebiete der Biologie. Protoplasma, 34: 238-313.

―――― 1941. Einiges über optische Anisotropie und Feinbau von Chromatin und Chromosomen. Chromosoma, 2: 86-111.

Schmitt, F. O. 1939. The ultra structure of protoplasmic constituents. Physiol. Rev., 19: 270-302.

―――― 1940. The molecular organization of protoplasmic constituents. Collecting Net, 15: 145.

―――― 1945. Ultrastructure and the problem of cellular organization. Harvey Lectures, 40: 249-268.

Schneider, B. 1933. Über die Umordunug der Chromosomen bei der Mitose. Z. Zellf., 17: 255-312.

―――― 1938. Die Zellteilung der Pflanzenzelle im Reihenbild (Beobachtungen an *Tradescantia virginica*). Z. Zellf., 28: 829-860.

Schrader, F. 1923. A study of the chromosomes in three species of Pseudococcus. Arch. Zellf., 17: 45-62.

―――― 1929. Experimental and cytological investigations of the life cycle of *Gossyparia spuria* (Coccidae) and their bearing on the problem of haploidy in males. Z. wiss. Zool., 134: 149-179.

―――― 1931. The chromosome cycle of *Protortonia primitiva* (Coccidae) and a consideration of the meiotic division apparatus in the male. Z. wiss. Zool., 138: 386-409.

―――― 1932. Recent hypotheses on the structure of spindles in the light of certain observations in Hemiptera. Z. wiss. Zool., 142: 520-540.

―――― 1934. On the reality of spindle fibers. Biol. Bull., 67: 519-534.

―――― 1935. Notes on the mitotic behavior of long chromosomes. Cytologia, 6: 422-431.

―――― 1936. The kinetochore or spindlefiber locus in *Amphiuma tridactylum.* Biol. Bull., 70: 484-498.

———— 1939. The structure of the kinetochore at meiosis. Chromosoma, 1: 230-237.

———— 1940a. The formation of tetrads and the meiotic mitoses in the male of *Rhytidolomia senilis* Say (Hemiptera Heteroptera). J. Morph., 67: 123-142.

———— 1940b. Touch-and-go pairing in chromosomes. Proc. Nat. Ac. Sci., 26: 634-636.

———— 1941a. The spermatogenesis of the earwig *Anisolabis maritima* Bon. with reference to the mechanism of chromosomal movement. J. Morph., 68: 123-148.

———— 1941b. Chromatin bridges and irregularity of mitotic coordination in the pentatomid *Peromatus notatus* Am. and Serv. Biol. Bull., 81: 149-162.

———— 1941c. The sex chromosomes: Heteropycnosis and its bearing on some general questions of chromosome behavior. In Cytology, Genetics and Evolution, Univ. Pa. Bicent. Conf.,: 27-37. Philadelphia, Univ. Pa. Press.

———— 1941d. Heteropycnosis and non-homologous association of chromosomes in *Edessa irrorata* (Hemiptera Heteroptera). J. Morph., 69: 587-608.

———— 1944. Mitosis; the movements of chromosomes in cell division. 1st ed. New York, Columbia University Press.

———— 1946a. The elimination of chromosomes in the meiotic divisions of *Brachystethus rubromaculatus* Dallas. Biol. Bull., 90: 19-31.

———— 1946b. Autosomal elimination and preferential segregation in the harlequin lobe of certain Discocephalini (Hemiptera). Biol. Bull., 90: 265-290.

———— 1947a. Data contributing to an analysis of metaphase mechanics. Chromosoma, 3: 22-47.

———— 1947b. The role of the kinetochore in the evolution of the Heteroptera and Homoptera. Evolution, 1: 134-142.

———— 1951. A critique of recent hypotheses of mitosis. In Symposium of Cytology. East Lansing, Michigan State Coll. Press.

Schrader, F., and C. Leuchtenberger. 1949. Variation in the amount of desoxyribose nucleic acid in different tissues of Tradescantia. Proc. Nat. Ac. Sci., 35: 464-468.

———— 1950. A cytochemical analysis of the functional interrelations of various cell structures in *Arvelius albopunctatus* (De Geer). Exp. Cell Res., 1: 421-452.

Schrader, F., and C. Leuchtenberger. 1952. The origin of certain nutritive substances in the eggs of Hemiptera. Exp. Cell Res., 3: 136-146.

Schreiner, A., and K. E. Schreiner. 1906. Neue Studien über die Chromatinreifung der Geschlechtszellen II. Arch. Biol., 22: 419-492.

Scott, A. C. 1936. Haploidy and aberrant spermatogenesis in a coleopteran, *Micromalthus debilis* Le Conte. J. Morph., 59: 485-515.

Seki, M. 1933. Zur physikalischen Chemie der histologischen Färbung. VIII, IX. Z. Zellf., 18: 1-56.

Sentein, P. 1950. Les transformations de l'appareil achromatique et des chromosomes dans les mitoses normales et les mitoses bloquées de l'oeuf en segmentation. Arch. Anat., 34: 377-394.

Sharp, L. W. 1934. Introduction to cytology. 3d ed. New York, McGraw-Hill.

Shimamura, T. 1940. Studies on the effect of the centrifugal force upon nuclear division. Cytologia, 11: 186-216.

Shinke, N. 1929. Chromosome arrangement. IV. Mem. Coll. Sci., Kyoto Univ., 4: 283-308.

Shiwago, P. I., and X. P. Troukhatchewa. 1940. Problème de la dynamique des "filaments de traction" dans la mitose. Arch. Anat. micr., 35: 457-464.

Sinnott, E. W., and R. Bloch. 1941. Division in vacuolate plant cells. Am. J. Bot., 28: 225-232.

Slack, H. D. 1938. The association of non-homologous chromosomes in Corixidae (Hemiptera-Heteroptera). Proc. Roy. Soc., Edinburgh, 58: 192-212.

Smith, F. H. 1935. Anomalous spindles in *Impatiens pallida*. Cytologia, 6: 165-176.

Smith, S. G. 1942. Polarization and progression in pairing. II. Canad. J. Res., 20: 221-229.

Spek, J. 1918. Oberflächenspannung als eine Ursache der Zellteilung. Arch. Entwmk., 44: 5-133.

Spooner, G. B. 1911. Embryological studies with the centrifuge. J. Exp. Zool., 10: 23-50.

Ssawostin, P. W. 1930. Magneto-physiologische Untersuchungen. I. Planta, 11: 683-726.

Staiger, H. 1950a. Chromosomenzahlen stenoglosser Prosobranchier. Experientia, 6: 54-59.

——— 1950b. Chromosomenzahl-Varianten bei *Purpurea lapillus*. Experientia, 6: 140-145.

Stedman, E., and E. Stedman. 1943. Chromosomin, a protein consti-

tuent of chromosomes. Nature, 152: 267-269.
——— 1947. The function of desoxyribose-nucleic acid in the cell nucleus. Soc. Exp. Biol. Symp., 1: 232-251.
Stern, C. 1931. Zytologisch-genetische Untersuchungen als Beweise für die Morgansche Theorie des Faktorenaustauschs. Biol. Zentralbl., 51: 547-587.
Storch, O. 1924. Die Eizellen der heterogenen Rädertiere. Zool. Jahrb. (Anat.), 45: 309-404.
Strangeways, T. S. P. 1923. Observations on the changes seen in living cells during growth and division. Proc. Roy. Soc., London, 94: 137-142.
Strasburger, E. 1875. Über Zellbildung und Zelltheilung. Jena, Fischer.
——— 1880. Über Zellbildung und Zelltheilung. 3d ed. Jena, Fischer.
Sturdivant, H. P. 1931. Central bodies in the sperm-forming divisions of Ascaris. Science, 73: 417-418.
——— 1934. Studies on the spermatocyte divisions in *Ascaris megalocephala*; with special reference to the central bodies, Golgi complex and mitochondria. J. Morph., 55: 435-475.
Sturtevant, A. H. 1936. Preferential segregation in triplo-IV females of *Drosophila melanogaster*. Genetics, 21: 444-466.
Suita, N. 1939. Studies on the male gametophyte in angiosperms. V. Jap. J. Genet., 15: 91-95.
Suomalainen, E. 1946. Die Chromosomenverhältnisse in der Spermatogenese einiger Blattarien. Ann. Acad. Sci. Fenn., 4: 4-60.
Swann, M. M. 1951a. Protoplasmic structure and mitosis I. J. Exp. Biol., 28: 417-433.
——— 1951b. Protoplasmic structure and mitosis II. J. Exp. Biol., 28: 434-444.
——— 1952. Structural agents in mitosis. In Internat. Rev. of Cyt., 1.
Swann, M. M., and J. M. Mitchison. 1950. Refinements in polarized light microscopy. J. Exp. Biol., 27: 226-237.
Swanson, C. P. 1942. Some considerations on the phenomenon of chiasma terminalization. Am. Nat., 76: 593-610.
Swanson, C. P., and R. Nelson. 1942. Spindle abnormalities in Mentha. Bot. Gaz., 104: 273-280.
Swift, H. H. 1950a. The desoxyribose nucleic acid content of animal nuclei. Physiol. Zool., 23: 169-198.
——— 1950b. The constancy of desoxyribose nucleic acid in plant nuclei. Proc. Nat. Ac. Sci., 36: 643-654.
Szent-Györgyi, A. 1951. Chemistry of muscular contraction. 2d ed.

New York, Academic Press.

Taylor, J. H. 1949. Increase in bivalent interlocking. J. Hered., 40: 65-69.

Tchou-Su. 1931. Etude cytologique sur l'hybridation chez les anoures. Arch. Anat. micr., 27: 1-106.

Teorell, T. 1937. Studies of the diffusion effect upon ionic distribution. J. Gen. Physiol., 21: 107-122.

Tharaldsen, C. E. 1926. The origin and nature of cleavage centers in echinoderm eggs. J. Exp. Zool., 44: 159-218.

Tischler, G. 1922. Allgemeine Pflanzenkaryologie. Berlin, Borntraeger.

Tjio, J. H. 1948. Notes on nucleolar conditions in *Ceiba pentandra*. Hereditas, 34: 204-208.

Tjio, J. H., and A. Levan. 1950. Quadruple structure of the centromere. Nature, 165: 368.

Trankowsky, D. A. 1930. "Leitkörperchen" der Chromosomen bei einigen Angiospermen. Z. Zellf., 10: 736-744.

Upcott, M. 1936. The mechanics of mitosis in the pollen-tube of Tulipa. Proc. Roy. Soc., London, 121: 207-220.

——— 1937a. The external mechanics of the chromosomes. VI. Proc. Roy. Soc., London, 124: 336-361.

——— 1937b. Timing unbalance at meiosis in the pollen-sterile *Lathyrus odoratus*. Cytologia, Fujii Vol.: 299-310.

Van Beneden, E. See Beneden, E. van.

Vandel, A. 1928. La parthénogenèse géographique. Bull. biol. France et Belg., 62: 164-284.

Vilkomerson, H. 1950. The unusual meiotic behavior of *Elymus wiegandii*. Exp. Cell Res., 1: 534-542.

Vincent, W. S. 1952. The isolation and chemical properties of the nucleoli of starfish oocytes. Proc. Nat. Ac. Sci., 38: 139-144.

Wada, B. 1935. Mikrurgische Untersuchungen lebender Zellen in der Teilung. II. Cytologia, 6: 381-406.

——— 1936a. Mikrurgische Untersuchungen lebender Zellen in der Teilung. III. Cytologia, 7: 198-212.

——— 1936b. Mikrurgische Untersuchungen lebender Zellen in der Teilung. IV. Cytologia, 7: 363-370.

——— 1938a. Experimentelle Untersuchungen lebender Zellen in der Teilung. I. Cytologia, 9: 97-109.

——— 1938b. Experimentelle Untersuchungen lebender Zellen in der Teilung. II. Cytologia, 9: 110-119.

——— 1939a. Experimentelle Untersuchungen lebender Zellen in der

Teilung. III. Cytologia, 9: 460–479.

——— 1939b. Experimentelle Untersuchungen lebender Zellen in der Teilung. IV. Cytologia, 10: 158–180.

——— 1940a. Lebendbeobachtung über die Einwirkung des Colchicins auf die Mitose, insbesondere über die Frage der Spindelfigur. Cytologia, 11: 93–117.

——— 1940b. On the spindle figures of the somatic mitosis in the prothallium cells of *Osmunda japonica* Thunb. in vivo. Bot. Mag. (Tokyo), 54: 89–95.

——— 1941. Über die Spindelfigur bei der somatischen Mitose der Prothalliumzellen von *Osmunda japonica* Thunb. in vivo. Cytologia, 11: 353–369.

——— 1949a. Studies on the mechanism of mitosis in the living state (in Japanese). Jap. J. Genet., 24: 51–61.

——— 1949b. Further studies on the effect of colchicine upon the mitosis of the stamen-hair in Tradescantia. Cytologia, 15: 88–95.

——— 1950. The mechanism of mitosis based on studies of the submicroscopic structure and of the living state of the Tradescantia cell. Cytologia, 16: 1–26.

Wald, H. 1936. Cytological studies on the abnormal development of the eggs of the claret mutant type of *Drosophila simulans*. Genetics, 21: 264–281.

Walton, A. C. 1924. Studies on nematode gametogenesis. Z. Zellf. Gewebe., 1: 167–239.

Warters, M., and A. B. Griffen. 1950. The telomeres of Drosophila J. Hered., 41: 183–190.

Wassermann, F. 1926. Zur Analyse der mitotischen Kern- und Zellteilung. Z. Anat., 80: 344–432.

——— 1929. Wachstum und Vermehrung der lebendigen Masse. In Handbuch der mikroskopischen Anatomie des Menschen. Berlin. Springer.

——— 1939. Mechanismus der Mitose. Arch. exp. Zellf., 22: 238–257.

Watase, S. 1891. Studies on cephalopods. I. J. Morph., 4: 247–303.

Weber, F. 1940. Eiweissspindeln von Valerianella. Protoplasma, 34: 148–152.

Wenrich, D. H. 1916. The spermatogenesis of *Phrynotettix magnus* with special reference to synapsis and the individuality of chromosomes. Bull. Mus. Zool., Harvard, 60: 57–134.

White, M. J. D. 1935. The effects of X-rays on mitosis in the spermatogonial divisions of *Locusta migratoria* L. Proc. Roy. Soc., London,

119: 61-85.
White, M. J. D. 1936. Chromosome cycle of *Ascaris megalocephala*. Nature, 137: 783.

———— 1937. The effect of X-rays on the first meiotic division in three species of Orthoptera. Proc. Roy. Soc., London, 124: 183-197.

———— 1938. A new and anomalous type of meiosis in a mantid, *Callimantis antillarum* Saussure. Proc. Roy. Soc., London, 125: 516-523.

———— 1940. The origin and evolution of multiple sex-chromosome mechanisms. J. Genet., 40: 303-336.

———— 1941. The evolution of the sex chromosomes. I. J. Genet., 42: 143-172.

Wilson, E. B. 1901. Experimental studies in cytology. I. Arch. Entwmk., 12: 531-596.

———— 1925. The cell in development and heredity. 3d ed. New York, Macmillan.

———— 1930. The question of the central bodies. Science, 71: 661.

———— 1931. The distribution of sperm-forming materials in scorpions. J. Morph., 52: 429-485.

———— 1932. Polyploidy and metaphase patterns. J. Morph., 53: 443-471.

Wilson, E. B., and A. F. Huettner. 1931. The central bodies again. Science, 73: 447.

Woker, G. 1920. Zur Physik der Zellteilung. Z. allg. Physiol., 18: 39-58.

Wolf, E. 1941. Die Chromosomen in der Spermatogenese einiger Nematoceren. Chromosoma, 2: 192-246.

———— 1950. Die Chromosomen in der Spermatogenese der Dipteren Phryne und Mycetobia. Chromosoma, 4: 148-204.

Wyckoff, R. W. G. 1934. Ultraviolet microscopy as a means of studying cell structure. Symp. Quant. Biol., 2: 39-47.

Yamaha, G. 1935. Über die pH-Schwankung in der sich teilenden Pollenmutterzelle einiger Pflanzen. Cytologia, 6: 523-527.

———— 1936. Weitere Beiträge zur Kenntniss über den isoelektrischen Punkt pflanzlicher Protoplasten. Sci. Rep. Tokyo Bunrik. Daig., 2.

Yamaha, G., and T. Ishii. 1932. Über die Ionenwirkung auf die Chromosomen der Pollenmutterzellen von *Tradescantia reflexa*. Cytologia, 3: 333-336.

Yamaha, G., and Y. Sinoto. 1925. On the behavior of the nucleolus in the somatic mitosis of higher plants, with microchemical notes. Bot. Mag. (Tokyo), 39: 205-227.

Yatsu, N. 1905. The formation of the centrosome in enucleate egg fragments. J. Exp. Zool., 2: 287-313.

Zeidler, J. 1925. Beiträge zur Frage des Galvanotropismus der Wurzeln. Bot. Arch., 9: 157-193.

Zeuthen, E. 1951. Segmentation, nuclear growth and cytoplasmic storage in eggs of Echinoderms and Amphibia. Pubbl. Staz. Zool. Napoli 23, Suppl. 47-69.

Zirkle, C. 1928. Nucleolus in root tip mitosis in *Zea mays*. Bot. Gaz., 86: 402-418.

―――― 1931. Nucleoli of the root-tip and cambium of *Pinus strobus*. Cytologia, 2: 85-105.

# Index

Abelson, P. H., and W. R. Duryee, 119
Acaridae, 32
Acroschismus, 22, 24, 40, 105
Actinosphaerium, 76
Aenoplex, 22
Agopanthus, 118
Akinetic chromosomes, 30; movement, 96
Alberti, W., and G. Politzer, 45
Alfert, M., A. W. Pollister, H. Swift, and, 53
Alga, 83, 90
Allium, 26, 27
Allolobophora, 7
Amphiaster, 38, 109, 125; appearance, 84, 98
Amphibia, 13, 26, 27, 32, 104, 119
Amphiuma, 26, 29, 31, 110
Anaphase, diagrammatic representation, 9
Anaphase movement of chromosomes, 58, 70 ff., 77; absolute velocity, 55; interpretations, 70 ff.; two processes, 79
Andrews, F. M., 16, 46
Angiosperms, 22
Anisolabis, 25, 60, 61, 62, 77, 78, 80
Aphididae, 45, 77, 113
Arion, 37, 38
Arrhenatherum, 55
Artemia, 41, 80
Artifacts, spindle fibers regarded as, 8, 13, 15, 21
Arvelius, size of spindle, 51, 52
Ascaris, 21, 22, 24, 31, 32
Asplanchna, 23$n$, 36
Astbury, W. T., 72
Aster, 23$n$; in live cells, 7, 12, 20, 22, 43; structure, 34-35; similarity to continuous fibers, 36; in cells recovering from hydrostatic pressure, 90
Astral rays, 39, 84
Attraction, 84-88 *passim*, 98, 114, 115 ff., 118
Atwood, K. C., T. Hinton and, 116
Aubert, J., R. Matthey and, 69

Autosomes, 61; lateral displacement of, 67
Autonomy, chromosomal, 106

"Balance theory of mitosis," 88
Baltzer, F., 105
Banga, I., and A. Szent-Györgyi, 34
Barber, H. N., 55, 82, 83, 93
—— and H. G. Callan, 102
Bauer, H., 109, 124
Beadle, G. W., 124
Beams, H. W., and R. K. King, 16
Beams, H. W., and T. C. Evans, 57
Beaumont, J. de, A. Naville and 116
Beck, L. V., and H. Shapiro, 119
Becker, W. A., 36, 44, 96, 114
Belar, K., 6, 14, 16, 17, 21, 26, 40, 65, 66, 83, 92 f., 96, 109, 110, 114; brief for longitudinal structure in living spindle, 15; mitotic hypothesis, 76-79 *passim*
—— and W. Huth, 77, 109
Bellevallia, 26
Belling, J., 24, 109
—— and A. F. Blakeslee, 24
Beneden, E. van, 71
—— and A. Neyt, 24
Benoit, J., and R. Kehl, 27
Bensley, R. R., 34
—— and N. L. Hoerr, 34
Berger, C. A., 109
Bernal, J. D., 101, 102
—— and I. Fankuchen, 101
Bernstein, J., 84
Berry, R. O., 120
Bersa, E., and F. Weber, 89
Berthold, G., 93
Biochemical methods, 47 ff.
Bipolarity, establishment of, 59, 60, 62
Birefringence, 17-20, 73, 101; in living asters, 34, 35; absence of, in interzonal region, 43
Bivalents, pre-metaphase stretch, 120; chromosomal fibers in, 121
Bjerknes, V., 97-100

# INDEX

Blakeslee, A. F., J. Belling and, 24
Blattidae, 122
Bleier, H., 8, 9, 16, 36, 44, 79, 81, 82, 83, 110, 111, 123, 124
Bloch, R., E. W. Sinnott and, 46, 47
Body repulsion, 88
Bonnevie, K., 16, 77
Boreus, 29, 122
Botta, B., 89
Bouin fixation, 14
Bouquet stage, 25, 60, 116
Boveri, T., 5, 21, 24, 71, 109
Bowen, R. H., P. Frew and, 95, 96
Brachystethus, 61, 66, 67
Breeding experiment, 123
Bresslau, E., 21
Brieger, F., 78
Brownian movement, 16, 96
Bryan, J. H. D., 48
Bucciante, L., 54
Bufo, 26
Bütschli, O., 6, 97

Callan, H. G., 119
—— H. N. Barber and, 102
—— J. T. Randall, and S. G. Tomlin, 119
Camara, A., N. Malheiros, D. de Castro, and, 32
Cannon, H. G., 86, 97, 99
Carlson, J. G., 7, 12, 14, 30, 35, 44
Carothers, E. E., 29
Caspersson, T., 48, 49, 51
Castro, D. de, 32
—— A. Camara, N. Malheiros, and, 32
Cataphoresis, 89, 91
Cells, fixed: structure, 8-12
Cells, living, 13, 17, 20, 22, 33, 100; orderly distribution of chromosomes to new, 4; structure, 6-8; division, x, 4, 37; chemistry, 47-53, 90, 124; division rate: agency that affects certain structures or activities, 56 f.; to-and-fro movement of chromosomes in, 66; expansion and contraction of localized regions in, 83; passing of a current through, 89
Cell walls, distortion of, 77, 80
Center, definition, 10; role in mitotic cycle, 20-25 *passim;* diffuse, 22; attraction between kinetochore and, 34, 121; interaction between chromosomes and, 59-61; interaction between non-chromosomal elements and, 62-63; establishment of bipolarity, 62; origin, 84; center-kinetochore repulsions, 87 f.; pulsation, 97 ff.; oscillation, 98 ff.; *see also* Poles
Centralization, of forces in spindle, 20 ff.
Centrifuging, effect on chromosomal fibers, 16 f., 46; effects on interzonal connections, 17, 19, 46; on continuous fibers, 46
Centrioles, definition, 10; reality, 20, 21, 23; categories of granules reported as: structure, 21; diffuse, 22, 32; origin, 23, 24; kinship to kinetochore, 33; extra, 34; oscillation and pulsation, 98 ff.; in bouquet, 116
Centrogenes, *see* Micelles
Centromere, *see* Kinetochore
Centrosome, 34, 100
Cephalaria, 27
Cerebratulus egg, 24
Chaetopterus, 35, 43, 109
Chambers, R., 14, 34, 35, 42
—— and H. B. Fell, 119
Chargaff, E., D. Elson and, 50
Chemistry, 47-53, 90, 124
Chiasma, 115
Chondriosomes, 6, 12, 19
Chorthippus, 6, 15, 26, 40, 41, 92, 110
Chortophaga, 7, 80
"Chromatin," 49
Chromatin bridges, 45, 46
Chromatography, paper, 47
Chromatoid bodies, 68
Chromomeres, 29, 87, 115; terminal, 116; believed to represent the telomeres, 117
Chromonema, *see* Genonema
Chromosomal fibers, 9, 19; definition, 10; effect of centrifuging, 16 f., 46; relation with kinetochore, 29, 39, 64, 69, 73, 96; fibrillae, 32; structure, 39-43; origin, 39 ff.; connection between pole and kinetochore, 40; elongation, 66 f.; contraction and expansion (pulling and pushing action), 66, 67, 70-81; elasticity, 71, 72, 121; macroscopic concept of action of, 72; packing of oriented particles, 74; elongation of half-spindles at metaphase, 77; shortening of half-spindle, 80; in bivalents and univalents, 121

# INDEX

Chromosomes, independent of cytoplasm, x, 4 f.; orderly distribution of, to new cells, 4; visibility in living cells, 6; connected directly with the poles, 8 f., 40; connected with a continuous fiber, 9, 65; definition, 10; cytological conception of, 10 f.; structure, 11; miniature spindle formed by lost, 24; importance of kinetochore in movements of, 25; bouquet formation of leptotene or pachytene, 25, 116; multiple or "Sammel," 31; terminal regions or ends, 31, 60, 116-17; stretched to form a bridge, 45; proteins, 48, 49; DNA, 49-53; division: method and timing of restoration, 51, 53; influence of colchicine, 57; interaction between centers and, 59-61; maneuvers independent of centers, 61; placement in a flat plate, 64, 65; distribution of, with respect to each other, 67; electric charges, 68; shrinkage, 68; orientation, 70, 99; pairing of, 70, 87; mutual repulsion of bodies of, 88; autonomy, 106-12, 121; interchange of regions, 115; see also Attraction; Hypotheses of mitosis; Repulsion
Churney, L., 119
—— and H. M. Klein, 89
Ciona, 36
Clark, F. J., 23, 124
Claude, A., and T. S. Potter, 50
Cleavage furrow, 47
Cleland, R. E., 111
Cleveland, R. L., 18, 21, 27, 31, 38, 39, 104, 105
Coagulation, 17
Coccidae, 7, 12, 40, 44, 45, 108 passim
Colchicine, 57
Coleman, L. C., 27
"Collochores," 116
Combination spindles, 9
Commissure, 26, 31
Conard, A., 46, 47, 83
Conklin, E. G., 35
Continuous fiber, 19; origin, 9, 37; definition, 10; structure, 36-39; in live cells, 37, 38; optical properties, 38; extranuclear, 38; result of polar activity, 39; connection with kinetochore, 41; relation to interzonal connections, 43 ff.; in anaphase, 46; changes in, while passing from metaphase to anaphase, 47; failure to appear in metaphase, 47; sliding action, 66, 76, 79; expansion, 79
Contraction and elongation: pulling action, 70-75; and expansion 76-81, 83, 100, 123; by dehydration, 83
Cooper, K. W., 17, 29, 33, 38, 70, 111, 116, 122
Copepoda, 36
Cornman, I., 72
—— and M. E. Cornman, 57
Creighton, H. B., and B. McClintock, 115
Crepidula, 36
Crepis, 26
Crustacea, 17, 41, 80, 120
Cucurbita, 95
Currents and streams, 93-97
Cyclops, 41 (see Fig. 8)
Cytasters, 24, 37
Cytochemical analysis, 48, 51 ff.
Cytogenetic research, 8, 123
Cytoplasm, 34, 35; independence of chromosomes and, x, 4 f.; cleavage of, 24; visibility of fibers in, 38; nucleolus correlated with growth of, 51; influence of colchicine, 57; electropositive, 89; influence of magnetic forces, 90
Cytoplasmic currents, 93, 97

Dalcq, A., 37
Dan, K., D. Mazia and, 51
Darlington, C. D., 26, 27, 28, 29, 30, 33, 45, 53, 87, 88, 115, 116, 118
—— and P. T. Thomas, 23
Daugherty, K., L. V. Heilbrunn and, 89
Degagnya, 83
Dehydration, contraction due to, 83
Delbrück, M., 115
Desoxyribonucleic acid (DNA), 49-53
Devisé, R., 42
Diakinesis, spacing in, 68
Diffusion hypothesis, 90-93, 94, 123
Diptera, 61, 118 and passim
"Distance conjugation," 116
Division furrow, 47
DNA, see Desoxyribonucleic acid
Dobzhansky, T., 78
Dorsoventrality, 70
Drosera, 22
Drosophila, 21, 31, 36, 55, 100, 111, 117, 124

## INDEX

Drüner, L., 66, 77, 103
Duryee, W. R., P. H. Abelson and, 119

Echinodermata, 21, 23, 73, 105, 108, 119
Eigsti, O. J., 57
Elastic action, 71, 72
Elastic fibers, 91, 82
Electrical forces, 123, and *passim*
Electrical phenomena, diffusion, 90-93
Electric charges, 68, 89
Electric hypothesis, 65, 84-90
Electrostatics, 84-90
Ellenhorn, J., 45, 96
Ellerström, S., and J. H. Tjio, 31
Elongation of spindle, 77 ff., 92, 120
Elson, D., and E. Chargaff, 50
Endomitosis, 109
Enslin, O., H. Freundlich, K. Söllner and, 101
Ephrussi, B., 54
Equilibrium of repulsion and attraction, 87, 88
Erickson, R. O., M. Ogur, G. U. Rosen, K. B. Sax, C. Holden, and, 48
Ernst, R. A., B. J. Luyet and, 119
Evans, T. C., H. W. Beams and, 57
Expansion, pushing, 75 f.; contraction and, 76-81, 83, 123; physicochemical aspects of, 80, 83; by hydration, 83; and contraction, 100
Expansion hypothesis, 75 f., 82
Extra-chromosomal proteins, 49

Fabergé, A. C., 115
Fankhauser, G., 37
Fankuchen, I., J. D. Bernal and, 101
Fautrez-Firlefyn, N., L. Lison and, 120
Federley, H., 33
Fell, H. B., R. Chambers and, 119
Festuca-Lolium hybrid, 23
Fiber, definition, 12
Fibers, *see* Chromosomal fibers; Continuous fiber; Spindle
Fibrillae of chromosomal fiber, 29, 32, 41, 42
Fischer, A., 21
Fixing fluids, effects on spindle, 13; effect on center, 19
Flagellata, 18, 21, 27, 38 and *passim*
Flemming, W., 6, 49
Fol, H., 84
Foot, K., and E. C. Strobell, 7
Freundlich, H., 41, 101

—— O. Enslin, and K. Söllner, 101
Frew, P., and R. H. Bowen, 95, 96
Frey-Wyssling, A., 106
Friedrich-Freksa, H., 115
Fritillaria, 29
Frolowa, S. L., 118
Fry, H. J., 21, 24, 46, 47
Fujii, K., and K. Yasui, 72
Fürst, E., 20
Fusom, 46

Gallardo, A., 84
Galtonia, 27
Gay, H. V., B. P. Kaufmann, M. R. McDonald, and 50
Geitler, L., 10, 11, 88, 109, 115
Genetics, linked with problems of mitosis, 3, 23, 123
Genonema (or chromonema), 28, 29, 32, 49; definition, 11
Gersch, M., E. Ries and, 13
Gilson Carnoy, 20
Goldacre, R. J., 73
Gossyparia, 22, 108
Granules, 20; categories of, reported as centrioles, 21; at center of cytasters, 24; movement toward pole, 96
Grassé, P. P., 27
Grégoire, school of, 13
Griffen, A. B., and W. S. Stone, 31
Griffen, A. B., M. Warters and, 117
Gross, F., 41, 80

Haga, T., H. Matsuura and, 117
"Half-spindle," 10
Half-spindle component, *see* Chromosomal fibers
Hardy, W. B., 84
Hartig, T., 101
Hartog, M., 84
Harvey, E. B., 5, 37
Heiderich, F., 21
Heilbrunn, L. V., 34, 35, 89
—— and K. Daugherty, 89
Hemiptera, 31, 44, 79, 104, 116, 120 and *passim*
Heterochromatin, 118
Heteropycnotic attraction, 61, 116, 117-18
Hinton, T., 116
—— and K. C. Atwood, 116
Hirschler, J., 12, 20, 45, 46

## INDEX

Histone, 50, 51
Hoerr, N. L., R. R. Bensley and, 34
Holden, C., R. O. Erickson, M. Ogur, G. U. Rosen, K. B. Sax, and, 48
Holomastigotoides, 31
Homarus, 69
Hoppe, E. N., C. W. Metz, M. S. Moses, and, 106
Howard, A., and S. R. Pelc, 53
Huettner, A. F., 36
—— and M. Rabinowitz, 21, 100
—— E. B. Wilson and, 21
Hughes, A. F., 56, 67, 103, 106
—— and M. M. Swann, 18, 80
Hughes-Schrader, S., 7, 22, 33, 40, 45, 60, 67, 80, 105, 108, 109, 116, 120, 121, 122
—— and H. Ris, 12, 32
Huskins, C. L., 115
—— and S. G. Smith, 27
Huth, W., K. Belar and, 77, 109
Hyaloplasm, fluid: astral rays paths of, 34
Hybridization, 45, 123
Hydration, 123; viscosity and, 81-83
Hydrodynamic forces, hypothesis of, 97-100
Hydrostatic pressure, 5, 90
Hypertonic media, 15, 16, 17, 77
Hypotheses of mitosis, 54-112

Impatiens, 27, 77
Inamdar, N. B., 121
Inoué, S., 18, 20, 35, 46, 47, 73, 80
Interionic forces, 101, 106
Interzonal connections, definition, 10; effect of centrifuging, 17, 19; structures, 17, 19, 43-47; birefrigence, 43; derived from coating of chromosomes, 46
Interzonal region, fibrous structure in anaphase and telophase, 46; subject to elongation, 77; expansion, 79
Inversion bridges, 45
Iriki, S., O. Minouchi and, 26
*Isagoras subaquilus*, 122
Iwata, J., 26, 27

Jacobson, W., and M. Webb, 44, 50, 86
Jehle, H., 115
Johnson, H. H., 21, 62
Juel, H. O., 109

Kamiya, N., 89
Karyokinesis, x, 5, 57
Karyomeres, 18, 38
Kattermann, G., 33
Kaufmann, B. P., 31
—— M. R. McDonald, and H. V. Gay, 50
Kehl, R., 27
Kerr, W. E., M. M. Rhoades and, 33
Kinetochore (or centromere), 25-34; definition, 12; structure, 24 ff.; terminology, 25; size, 26; staining capacity, 26, 27; relation to chromosomal fibers, 29; fragmentation: misdivision, 30; cyclical alterations, 31; fragmentation by X rays, 32; multiple or polykinetic condition, 32; diffuse, 32, 33; interrelation between center and, 33, 59, 121; relation to spindle fibers, 39, 40; connection between continuous fiber and, 41; bipolar orientation, 64 f., 69 f.; secretory fluid from, 66; mutual interaction between two kinetochores, 70; responsible for loss of birefringence in spindle, 73; in electric hypothesis, 86, 87; repulsion, 88, 120; chromosomal fiber traced to, 96; division of spindle spherules, 98; autonomy, 109; attraction, 118; separation, 120; orientation of, reversible, 121
Kinetochore-center attraction, 121
Kinetonema, see Genonema
King, R. L., H. W. Beams and, 16
Kinosome, see Spindle spherules
Klein, E., 70
Klein, G., E. Klein, C. Leuchtenberger, and, 53
Klein, H. M., L. Churney and, 89
Klingstedt, H., 45, 123
Koerperich, J., 36, 110
Koller, P. C., 29, 87
Koonz, C. H., 22
Koslov, V. E., 26
Kossel, A., 47
Kupka, E., and F. Seelich, 41
Küster, E., 101
Kuwada, Y., 86

Lagerstedt, S., 51
Lamb, A. B., 97, 100
Lamellae, 15; flow of viscous protoplasm between, 94, 95, 96

# INDEX

Lamellar form of spindle fibers, 15, 94, 96
Lams, H., 37
Landau, E., 97
Lathyrus, 124
Lehotzky, P. von, 89
Lepidoptera, 21, 33 and *passim*
Leptotene stage, 60
Lettré, H., 57
—— and R. Lettré, 57
Leuchtenberger, C., F. Schrader and, 49, 51, 53, 120
Leuchtenberger, C., G. Klein, and E. Klein, 53
Leuchtenberger, C., and H. Z. Lund, 53
Leuchtenberger, C., R. Leuchtenberger, R. Vendrely, and C. Vendrely, 48
Levan, A., J. H. Tjio, and, 28
Lewis, M. R., 13, 15
Lilium, 17, 26, 27, 42, 73
Lillie, F. R., 5, 16, 46, 109
Lillie, R. S., 68, 89, 111; electrostatic hypothesis, 84 ff.
Lima-De-Faria, A., 27, 28
Linderstroem-Lang, K., 47
Liogryllus, 41
Liquid crystals, *see* Tactoids
Lison, L., and J. Pasteels, 53
Lison, L., and N. Fautrez-Firlefyn, 120
Living cells, *see* Cells, living
Llaveia, 22, 45, 80
Loligo, 75
Lorbeer, G., 116
Love, R. M., 31
Loxa, 63, 69
Lucas, F. F., and M. B. Stark, 12
Lund, H. Z., C. Leuchtenberger and, 53
Luyet, B. J., and R. A. Ernst, 119
Luzula, 32

McClendon, J. F., 97
McClintock, B., 26, 30, 45, 117
—— H. B. Creighton and, 115
McDonald, M. R., B. P. Kaufmann, H. V. Gay, and, 50
Magnetic hypothesis, 84-90
Magnetic or electrical forces, chromosomal mechanics on basis of, 84
Malheiros, N., D. deCastro, A. Camara, and, 32
Mantidae, 69, 109; meiotic prophase, 120
Mark, H., K. H. Meyer and, 72
Martens, P., 13, 54

Mathematical analysis of cellular phenomena, 91 ff.
Matrix, definition, 11; relation to interzonals, 45; nonbasic proteins, 49
Matsuura, H., 26, 28, 29, 116
—— and T. Haga, 117
Matthey, R., 122
—— and J. Aubert, 69
Mazia, D., 49, 50
—— and K. Dan, 51
Mecistorhinus, 61, 66, 67
Mecoptera, 122
Meiotic mitosis, 26; term, x, 5
Meiotic pairing, 115 f.
Melanoplus, 12
Melophagus, 111
Mesostoma, 21
Metaphase, 4; changes encountered in, 54; period prior to, 58 ff.; elastic chromosomal fibers responsible for, 93
Metaphase mechanics, 64-70
Metaphase plate, 14, 58, 91, 94, 98, 99, 103; placement of chomosomes with respect to each other, 64, 68; in electric hypothesis, 84, 86, 88
Metaphase spindle, 36
Metz, C. W., 22, 23, 68, 82, 104, 106, 107, 108, 110, 111
—— and J. F. Nonidez, 12
—— M. S. Moses, and E. N. Hoppe, 106
Metzner, R., 25, 27
Meves, F., 21, 37, 66, 77
Meyer, K. H., 72
—— and H. Mark, 72
Micelles (centrogenes), 28, 30, 35, 41, 101, 125
Michel, K., 66
Microdissection, 14
Micromalthus, 22, 23, 108
Micromethod, 47
Microscopy, new type, ix, 48
Midbodies, 46
Miescher, F. 47
Minouchi, O., and S. Iriki, 26
Mirsky, A. E., 34
—— A. W. Pollister and, 50
—— and H. Ris, 50
Mitosis, difficulties in way of solution, 3, 113, 123; genetics linked with problems of, 3, 23, 123; independence of elements in cycle, 4; definition, 5; part played by centers, 20-25; chemistry of

cell constituents, 47-53; hypotheses of, 54-112; interrelation of all steps in cycle: time duration, 54 ff.; experimental analysis, 56-58; period prior to metaphase, 58; prophase movements, 58-64; metaphase mechanics, 64-70; post-metaphase movements, 70 ff.; contraction: pulling, 70-75; expansion: pushing, 75 f., 80; variations: contraction and expansion, 76-81; viscosity and hydration, 81-83; hydrating and dehydrating agents, 83; electrostatics, 84-90; time duration, 88; diffusion hypothesis, 90; streaming: currents, 93-97; hydrodynamic forces, 97-100; tactoid hypothesis, 100-106; chromosome autonomy, 106-12; related problems, 113-22; dissociating each step into its component processes, 113; resting stage, 114; single type of force underlies all mitotic activity, 123; main points of attack, 124 f.
Mitotic processes, classification, 8
Möllendorf, W. von, 56, 83
Mollusca, 33, 37, 118, 122
Monasters, 108 f
Monné, L., 119
Monocotyledons, 33
Monopolar spindles, 104
Moore, S., 47
Morgan, T. H., 16, 46
Moses, M. S., C. W. Metz, E. N. Hoppe, and, 106
Mottier, D. M., 42
Muller, H. J., 116
Müntzing, A., R. Prakken and, 33
Muscle fibrils, *see* Myofibrils
Muscular action, 72
Myofibrils, 15, 72, 73

Najas, 26
Nautococcus, 22
Naville, A., and J. de Beaumont, 116
Nawaschin, S., 25, 31
Nebel, B. R., 26, 27, 28, 30, 115
Negative tactoids, 101, 102
Nelson, R., C. P. Swanson and, 124
Nematoda, 7, 31, 70
Nemec, B., 19
Neuroptera, 116
Neyt, A., E. van Beneden and, 24
Nishiyama, I., 30

Nonidez, J. F., C. W. Metz and, 12
Nuclear membrane, 25, 58, 62, 118; formation around telophase group, 4; chromosome movements prior to disappearance, 67; depolarized, 84; role in mitotic cycle, 118-20
Nuclear spindles, 7 ff., 36, 110
Nucleic acid, 49-53
Nucleolus, 51, 68, 86, 95, 96, 97, 119, 120
Nucleoproteins, 48; in cells, 47; droplets of, extruded through nuclear membrane, 120
Nucleus, chemistry, 47-53; interrelation of centers and, electropositive contents, 89

Odonata, 33
Oecanthus, 21
Oenothera, 69, 111
Ogur, M., 48
—— R. O. Erickson, G. U. Rosen, K. B. Sax, and C. Holden, 48
Oksala, T., 33
Oligopyrene cells, 30, 33
Oligopyrene spermatocytes, 118
Oöcyte, 119 and *passim*
Opisthacanthus, 27
Optical properties, of live spindle, 6, 15; of aster, 7, 43; of continuous fibers: of spindle elements, 43
Orthoptera, 21, 27, 30, 123 and *passim*
Oscillation and pulsation, 97-100
Östergren, G., 28, 57, 66-70 *passim*, 73, 74, 75, 103, 122
—— and R. Prakken, 33
Osterhout, W. J. V., 42

Pachylis, 19
Pachytene stage, 60
Painter, T. S., 23, 37
—— and K. Reindorp, 109
—— and W. Stone, 31
Pairing, meiotic, 115 f.
Paper chromatography, 47
Paragenoplastin, 9, 36, 110
Pasteels, J., L. Lison and, 53
Payne, F., 111
Pease, D. C., 5, 90
Pediculopsis, 13, 17, 18, 38
Pelc, S. R., A. Howard and, 53
Pellicle (or sheath), 44, 45, 110; definition, 11

*Perlodes microcephala*, 69
Pfeiffer, H. H., 18, 101
Phase-contrast microscope, ix, 119
Phasmida, 122
Phenacoccus, 22, 108
Phleum, 31
Phosphoprotein, 51
Phragmoplast, 47
Physical chemistry of the cell, 47-53, 90, 124
Physicochemical aspects of expansion, 80
Picea, 27
Pilawski, S., 20
Pinney, E., 27
Pinus, 86
Piza, S. de T., 33, 70, 73
Polar activity, continuous fiber and astral ray result of, 39
Polar granules, 27
Polarization, 22, 23, 24, 79, 84, 85, 124
Polarization microscope, 18
Polarized light, 18
Poles of spindle, 8, 87; definition, 10; centralization of forces at, 20 ff., 40; altered, 23; miniature, 24; continuous fibers and astral rays the result of polar activity, 37, 39; relation to chromosomal fibers, 39, 40; interaction between kinetochore and, 41, 69; multipolar origin of spindles which become bipolar, 43; movement of chromosomes to, 54, 74; distance between, at telophase, 80; electrostatic hypothesis reorigin, 84 ff.; repulsion-diffusion forces, 91; opposing currents, 94; poleward streaming and currents, 94 ff.; pulsating or oscillating, 98, 99, 100
Politzer, G., W. Alberti and, 45
Pollister, A. W., 21, 33, 34, 35, 48
—— and A. E. Mirsky, 50
—— and H. Ris, 48, 51
—— and P. F. Pollister, 30
—— H. Swift, and M. Alfert, 53
Positive tactoids, 101, 102
Post-metaphase movements, 70 ff.
Potter, T. S., A. Claude and, 50
Prakken, R., and A. Müntzing, 33
Prakken, R., G. Östergren and, 33
Pre-metaphase stretch, 120-22
Primary constriction, kinetochore known as, 26

Prometaphase, 39; give-and-take action during, 66
Propach, H., 27
Prophase, 4, 58 and *passim*
Prophase movements, 54, 58-64
Propulsion, 65, 91
Protein chains hypothesis, 72 ff.
Protein molecules, orientation of, 18
Proteins, in chromosomes, 48, 49; studies of, 101
Protoplasmic streaming, 90
Protortonia, 45
Protozoa, 21
Pseudococcus, 22, 108
Pulling, contraction and elongation, 70-75
Pulling hypothesis, 66 ff., 76
Pulsation and oscillation, 97-100
Pushing hypothesis, 75 f., 80, 124
Pushing mechanism, 65 ff.; "Stemmkörper," 76-79 *passim*, 83, 92

Rabbit, 100
Rabinowitz, M., 54
—— A. F. Huettner and, 21, 100
Rabl orientation, 60
Rana, 21
Randall, J. T., H. G. Callan, S. G. Tomlin and, 119
Rashevsky, N., 91-93, 66, 72
Refractive index, 6, 13, 20
Regemorter, D. van, 47
Reindorp, K., T. S. Painter and, 109
Repulsion, 62, 68, 84; mitotic activity based on, 85; and attraction, 87, 98; kinetochore, 87 f., 120; diffusion forces, 91; term, 114
Resting stage, 4, 114 and *passim*
Rhabditis, 7
Rhoades, M. M., 30
—— and H. Vilkomerson, 33
—— and W. E. Kerr, 33
Rhomaleum, 6
Rhytidolomia, 116
Ribonucleic acid (RNA), 48-53
Ribonucleoproteins, 48-53
Ries, E., and M. Gersch, 13
RNA, *see* Ribonucleic acid, 50
Ris, H., 7, 32, 55, 77, 79, 80, 113
—— A. E. Mirsky and, 50
—— A. W. Pollister, and, 48, 51
—— S. Hughes-Schrader and, 12, 32
Robertson, W. R. B., 21

# INDEX

Robyns, W., 13
Rosen, G. U., R. O. Erickson, M. Ogur, K. B. Sax, C. Holden, and, 48
Runnström, J., 18

Sakamura, T., 25
Salamander, 37
Salazar, A. L., 100
Sax, K., 124
Sax, K. B., R. O. Erickson, M. Ogur, G. U. Rosen, C. Holden, and, 48
Schaede, R., 16, 27, 94-97
Schmidt, W. J., 18, 34, 35, 43, 73
Schmitt, F. O., 14, 18
Schneider, B., 6
Schrader, F., 16-22 passim, 25-33 passim, 40-46 passim, 52, 60-63 passim, 66-73 passim, 78, 80, 96, 104, 105, 108, 110, 116, 124
——— and C. Leuchtenberger, 49, 51, 53, 120
Schreiner, A., and K. E. Schreiner, 60
Sciara, 22, 23, 82, 106-8, 111, 120
Scorpionidae, 33
Scorzonera, 27
Scott, A. C., 22, 23, 104, 108
Seelich, F., E. Kupka and, 41
Seki, M., 13
Sex chromosomes, 61, 67, 111, 116, 121 and passim
Sex trivalent, malorientation of, during stretch, 121
Shapiro, H., L. V. Beck and, 119
Sharp, L. W., 28, 29
Sheath, see Pellicle
Shimamura, T., 7, 16, 17, 46, 72
Shiwago, P. I., and X. P. Troukhatchewa, 72
Sinnott, E. W., and R. Bloch, 46, 47
Sinoto, Y., G. Yamaha and, 96
Smith, F., 77
Smith, S. G., C. L. Huskins and, 27
Söllner, K., H. Freundlich, O. Enslin, and, 101
Spacing in diakinesis, 68
Spek, J., 34, 97
Spermatocytes, 92, 122
Sperm heads, 50
Spherules, see Spindle spherules
Spindle, rigidity, 7; nuclear, 7 ff., 36, 110; fibers regarded as artifacts, 8, 13, 15; direct type, 8, 9, 40; gathering of, for mass analysis possible, 8$n$; poles ($q.v.$), 8, 87; combination, 9; indirect type, 9, 40; components of various types, 10; lack of structure in the living, 12; actuality of structural elements, 12-20; lamellae, 15, 94, 96; Belar's observations, 15 f.; direct obesrvation and birefringence, 17-20; monocentric, 23; miniature, 24, 37, 109; spherule, 26, 28, 29, 98; constituents, 35 f.; internuclear, 36; origin, 36, 40, 41; structural elements, 36-47; continuous fibers ($q.v.$), 36-39; cytoplasmic, 36, 37, 40 (see Fig. 6); visibility of fibers in, 38; chromosomal fibers ($q.v.$), 39-43; multipolar, 42; interzonal connections ($q.v.$), 43-47; dimensions, 51, 52; effect of colchicine, 57; as a dynamic system, 68; loss of birefringence in anaphase, 73; pushing action, 76 basic framework, 76; elongation, 77 ff., 92; diffusion forces, 92; hypothesis involving streaming or currents in, 93 ff.; differences between tactoids and, 103; conditions during division in protozoa, 104; elongation, 120
Spindle spherules, 26, 28, 29, 98
Spooner, G. B., 16
Ssawostin, P. W., 90
Staiger, H., 122
Stainability of chromosomes, 49
Staining, methods, 8, 27; of nuclear structures, 119
Stark, M. B., F. F. Lucas and, 12
Stauroderus, 118
Steatococcus, 12, 32
Stedman, E., and E. Stedman, 49, 50
Stein, W. H., 47
"Stemmkörper" (pushing body), 76-79 passim, 83, 92
Stenobothrus, 93
Stern, C., 115
Stone, W., T. S. Painter and, 31
Stone, W. S., A. B. Griffen and, 31
Storch, O., 23$n$, 36
Strasburger, E., 6
Streaming hypothesis, currents, 93-97
Strobell, E. C., K. Foot and, 7
Structure, 6-53; living cells, 6-8; fixed cells, 8-12; actuality of structural elements in the spindle, 12; nature and origin of the spindle apparatus, 20-47; see also Spindle

Structure protein, 34
Sturdivant, H. P., 21
Sturtevant, A. H., 111
Substance parachromosomique, 36
Suita, N., 109
Suomalainen, E., 122
Surface tension, 68, 97
Swann, M. M., 41, 73, 106
—— A. F. Hughes and, 18, 80
Swanson, C. P., 109
—— and R. Nelson, 124
Swelling, *see* Expansion
Swift, H. H., 53
—— A. W. Pollister, M. Alfert and, 53
Synapsis, 61, 115, 117
Szent-Györgyi, A., 72
—— I. Banga and, 34

Tactoids, 41, 68, 100-106
Tamalia, 32, 113
Taylor, J. H., 117
Telomere, 45, 60, 116-17
Telophase, 4 and *passim*
Teorell, T., 90, 91
Terminal adhesion and attraction, 116
"Terminal affinity," 116
"Terminal chiasmata," 116
Terminology, 10 ff.
Tharaldsen, C. E., 24
Thomas, P. T., C. D. Darlington and, 23
Timing of mitosis, 54 ff., 88, 115, 124
Tischler, G., 93
Tissue cultures, 80; experiments on, 15
Tityus, 33
Tjio, J. H., and A. Levan, 28
Tjio, J. H., S. Ellerström and, 31
Tomlin, S. G., H. G. Callan, J. T. Randall, and, 119
Torpedo, 36
Traction fiber, 65 f., 76, 77, 79
Traction hypothesis, 70-75, 124
Tradescantia, 26, 29, 96, 117
Trankowsky, D. A., 26
Trillium, 26, 27, 29, 117
Trimerotropis, 6
Triton, 37
Troukhatchewa, X. P., P. I. Shiwago and, 72

Tulipa, 29

Ultraviolet light, 48
Univalents, 106, 111; chromosomal fibers in, 121
Upcott, M., 12, 29, 44, 87, 124
Urechis, 90, 109

Vanadium pentoxide, 101
Vandel, A., 110, 124
Velocity of chromosomes, 82
Vendrely, R., C. Vendrely, C. R. Leuchtenberger and, 48
Vilkomerson, H., 33
—— M. M. Rhoades and, 33
Vincent, W. S., 51
Viscosity, 65, 94 ff.; and hydration, 81-83

Wada, B., 36, 38, 41, 56, 65, 66, 74, 75
Wald, H., 124
Walton, A. C., 70
Warters, M., and A. B. Griffen, 117
Wassermann, F., 3, 8, 19, 34, 44, 66, 81, 82, 83
Watase, S., 75, 76
Webb, M., W. Jacobson and, 44, 50, 86
Weber, F., 101
—— E. Bersa and, 89
White, M. J. D., 26, 31, 45, 120, 124
Wilson, E. B., 5, 10, 21, 27, 41, 84, 109, 116
—— and A. F. Huettner, 21
Woker, G., 97
Wolf, E., 122
Wyckoff, R. W. G., 12

Yamaha, G., 6
—— and Y. Sinoto, 96
Yasui, K., K. Fujii and, 72
Yatsu, N., 24

Zea, 23, 26, 124; misdivision, 30
Zeidler, J., 89
Zeuthen, E., 50, 58
Zirkle, C., 86, 95, 96

Bei Fragen zur Produktsicherheit wenden Sie sich bitte an:
If you have any questions regarding product safety,
please contact:

Walter de Gruyter GmbH
Genthiner Straße 13
10785 Berlin
productsafety@degruyterbrill.com